Advanced Membrane Separation for Sustainability 2013

Topical Conference at the 2013 AIChE Annual Meeting: Global Challenges for Engineering a Sustainable Future

San Francisco, California, USA
3-8 November 2013

ISBN: 978-1-63439-030-9

Printed from e-media with permission by:

Curran Associates, Inc.
57 Morehouse Lane
Red Hook, NY 12571

Some format issues inherent in the e-media version may also appear in this print version.

Printed by Curran Associates, Inc. (2014)

AIChE
3 Park Avenue
New York, NY 10016-5991

Phone: (203) 702-7660
Fax: (203) 775-5177

www.aiche.org

Additional copies of this publication are available from:

Curran Associates, Inc.
57 Morehouse Lane
Red Hook, NY 12571 USA
Phone: 845-758-0400
Fax: 845-758-2634
Email: curran@proceedings.com
Web: www.proceedings.com

TABLE OF CONTENTS

(226e) A Sustainable Chemical Industry May Imply Dispersed Manufacturing

Cussler, E. L., University of Minnesota
Tiffany, D., University of Minnesota
Originally presented on: 11/4/2013 12:30:00 - 13:00:00

A SUSTAINABLE CHEMICAL INDUSTRY MAY IMPLY DISPERSED MANUFACTURING

E. L. Cussler and Douglas G. Tiffany

Departments of Chemical Engineering and Materials Science, and of

Agricultural Economics

University of Minnesota, Minneapolis MN 55455

A "sustainable chemical industry" implies using renewable feeds, especially those derived from agriculture. The ease of doing so depends on the amount of chemical produced. For example, antibiotics like penicillin already use agricultural waste products as a feed. The amount of antibiotic produced is tiny relative to commodity chemicals like olefins, and the transportation cost of transporting this high value-added product to market is minor. Thus the production of antibiotics will remain centralized, as it is now.

In contrast, the production of fuels from biomass will involve huge amount of feed and of product of much less value per kilo than antibiotics. Transportation costs of moving both feed and fuel will be major. As a result, the production of fuels will probably involve many smaller plants, whose lower efficiency is balanced by lower costs of transportation.

Thus a sustainable chemical industry may involve dispersed manufacture, rather than the highly centralized plants characteristic of today's industry. This paper will discuss such a future for ammonia, the key chemical fertilizer, which can be made from air, water, and wind power. We will describe what such a chemical plant could look like, and the advantages and debits of such dispersed manufacture.

(142d) Advanced Polymeric Membranes With Tailored Surface Properties

Ulbricht, M., Universitat Duisburg-Essen
Birkner, M., Universitat Duisburg-Essen
Frost, S., Universitat Duisburg-Essen
Kumar, M.,
Li, M.,
Quilitzsch, M.,
Pieper, T.,
Stahra, N.,
Sulc, R.,
Originally presented on: 11/4/2013 13:00:00 - 13:30:00

Advanced polymeric membranes with tailored surface properties

Mathias Ulbricht, Marc Birkner, Sven Frost, Mahendra Kumar, Miao Li, Mathias Quilitzsch, Thorsten Pieper, Nico Stahra, Robert Sulc

Lehrstuhl für Technische Chemie II, Universität Duisburg-Essen, 45117 Essen, Germany; mathias.ulbricht@uni-essen.de

Membrane technologies have been established in a wide range of industrial processes. However, in advanced separation and reaction engineering, membranes applied as separator or contactor can offer yet many more distinct advantages [1].

The majority of synthetic membranes are made from polymers because barrier and surface properties can be varied in wide ranges with help of established scalable manufacturing processes. Significant efforts have been made to further improve membrane performance by focusing on barrier properties (high selectivity, high flux) or surface properties (affinity, antifouling). Membrane fouling is still a major obstacle because it reduces productivity and product quality and increases energy consumption and need of additional chemicals for cleaning; hence, efficiency and sustainability of membrane processes are reduced.

In this presentation, we will discuss how tailored surface functionalizations can significantly improve the performance of established membranes and can also contribute to the development of membranes using novel separation principles. With respect to the preparation/manufacturing, two general routes will be outlined, *i)* membrane formation from functional copolymers or blends of established membrane polymers with functional copolymer additives, and *ii)* post-functionalization of membranes by (reactive) coating or grafting methods (cf. [2,3]). Important recent trends in the field will be illustrated with examples from own research.

One focus will be on ultra- and microfiltration membranes with hydrophilic, low-fouling surfaces. Such membranes can be obtained by phase separation of polymer solutions (route *i)*), either using tailored block copolymer additives (e.g. [4,5]) to established membrane polymers or *de novo* synthesized copolymers [6]. If the hydrophilic functional blocks comprise also groups which can reversibly bind certain molecules or particles, the resulting membranes can be used for combined separations using both size-selective barrier and adsorber surface properties. Another combined separation mechanism is the charge-enhanced selectivity in ultrafiltration, for instance of proteins [6].

With respect to route *ii)*, we will discuss recent progress toward ultra- and nanofiltration membranes with reduced fouling and improved cleaning. Thin-film hydrogel composite membranes obtained by graft copolymerization of hydrogels initiated at the membrane surface had been tailored with respect to the important parameters selectivity, flux and antifouling, either for selective separation of proteins and other nanoparticles [7] or for fractionation of smaller molecules and ions [8]. With zwitterionic grafted polymers and their stimuli-responsive properties it is also possible to develop more active antifouling and cleaning strategies, for instance based on a reversible conformation of the grafted "protective" layer change between operation and cleaning conditions. Finally, we will demonstrate that well-controlled grafting of a thermo-responsive polymer from the surface of ultrafiltration membrane pores can yield membranes where the size-selectivity can be tuned by the reaction conditions and the fractionation of nanoparticles can be reversibly switched by temperature.

All these strategies have relevance to widen the scope of the membrane applications and thus to contribute to the implementation or improvement of sustainable technologies.

[1] E. Drioli, L. Giorno (Ed.), *Membrane operations: Innovative separations and transformations*, Wiley–VCH, Weinheim, **2009**.

[2] M. Ulbricht, *Polymer* **2006**, *47*, 2217.

[3] Q. Yang, N. Adrus, F. Tomicki, M. Ulbricht, *J. Mater. Chem.* **2011**, *21*, 2783.

[4] H. Susanto, N. Stahra, M. Ulbricht, *J. Membr. Sci.* **2009**, *342*, 153.

[5] E. Berndt, M. Ulbricht, *Polymer* **2009**, *50*, 5181.

[6] M. Kumar, M. Ulbricht, *RCS Advances* **2013**, *accepted*.

[7] P. D. Peeva, T. Knoche, T. Pieper, M. Ulbricht, *Ind. Eng. Chem. Res.* **2012**, *51*, 7231.

[8] R. Bernstein, E. Antón, M. Ulbricht, *ACS Appl. Mater. Interf.* **20**

(142b) New Material Development for Membrane Separations

Noble, R. D., University of Colorado

Originally presented on: 11/4/2013 13:30:00 - 14:00:00

Synthetic membranes are important for many separation applications, such as the separation and purification of atmospheric gases for medical and industrial use, and salt removal from seawater in regions lacking fresh water sources. Unlike mesh filters used to separate large solid particles from a fluid (phase difference), membranes used to separate molecular species from one another (i.e., a chemical separation), is a much more difficult process [given the size scale and level of mixing of the components].

Because of the benefits of membrane-based chemical separations over other methods (e.g., smaller device footprints, lower energy use compared to distillation), there has been recent interest in applying membranes to new separation problems, as well as improving membrane materials used in existing separations. Consequently, there is a need for better membrane materials in general (i.e., with better selectivity, better productivity, longer life and operational stability, increased operating temperature range, use in chemically challenging environments, and the ability to adjust materials properties and performance for a given application).

The new materials are room temperature ionic liquids (RTILs), zeolites and lyotropic liquid crystals (LLCs). Each of these materials has unique physiochemical properties that provide an advantage in certain applications. Each of these materials will be discussed and their use in gas and water treatment.

For researchers working on the design of new membrane materials, it is important to not only consider the design factors that afford better separation selectivity but to also consider the other factors that afford good productivity as well. Often it is these materials engineering factors that decide whether a membrane material is viable in practical separation or not.

(142c) Hybrid Membranes for Improved Molecular Separation in Gas and Liquid-Phase

Tsuru, T., Hiroshima University

Originally presented on: 11/4/2013 14:00:00 - 14:30:00

Hybrid membranes have been investigated for improved performance which cannot be achieved by using only one type of materials. In this talk, our recent progresses on hybrid membranes will be introduced, including organosilica where Si was hybrid with organic functional groups on the molecular level, and mixed-matrix-membranes where inorganic nanoparticles were hybrid with polyamide during interfacial polymerization.

Amorphous silica membranes show highly permselectivity for hydrogen; however, there still remain two technical challenges: pore size control and hydrothermal stability.[1)] In order to improve hydrothermal stability of silica membranes, Castricum et al.[2)] developed hybrid silica membranes using bis (triethoxysilyl)ethane (BTESE), which showed excellent hydrothermal stability in pervaporation of aqueous alcohol solutions because of incorporated organic linking groups (Si-C-C-Si unit) in silica networks. We recently proposed a new strategy to control pore sizes of silica membranes using bridged alkoxides.[3-8)] Bis (tri-ethoxysilyl) ethane (BTESE), consisting of ethane group between 2 silicon atoms ($-Si-CH_2-CH_2-Si-$) as the minimum unit in hydrolysis and condensation reaction, was found to lead to loose silica networks. BTESE silica membranes showed approximately one order magnitude high H_2 permeance ($\sim 10^{-5}$ mol/(m^2 s Pa)) compared with previously reported silica membranes using TEOS (tetraethoxysilane), and a high H_2 to SF_6 permeance ratio of more than 1,000 with a low H_2 to N_2 selectivity (~10).[3)] Applications, including gas separation (hydrogen, organic gas[3-6)]) and liquid phase separation (reverse osmosis[7)], pervaporation[8)]), will be also discussed. BTESE membrane showed quite high rejection for NaCl and high stability to chlorine in reverse osmosis.

Regarding polyamide (PA) membranes for reverse osmosis, we proposed spray-assisted, 2-step interfacial polymerization[9)]. We found the formation of multilayered ridge-and-valley structures via the first spray-assisted interfacial polymerization, leading enhanced water permeation due to increased surface area. Zeolite Y nanoparticles, which were embedded into PA layer in the first-spray process, were confirmed to increase water permeability.

References

1) T. Tsuru, *J. Sol-Gel Sci. Technol.*, 46 (2008) 349; *J. Japan Petroleum Institute* 54 (2011) 277. 2) H.L. Castricum, A. Sah, R. Kreiter, D.H.A. Blank, J.F. Vente, J.E. ten Elshof,, *J. Mater. Chem. 18 (*2008) 2150. 3) M. Kanezashi, K. Yada, T. Yoshioka, T. Tsuru, *J. Am. Chem. Soc.* 131 (2009) 414. 4) H. R. Lee, M. Kanezashi, T. Yoshioka, T. Tsuru, *AIChE J.* 57 (2011) 2755. 5) G. Li, M. Kanezashi, T. Tsuru, *J. Membr. Sci.* 379 (2011) 287. 6) M. Kanezashi, M. Kawano, T. Yoshioka, T. Tsuru, *Ind. Eng. Chem. Res.* 51 (2012) 944. 7) R. Xu, J. Wang, M. Kanezashi, T. Yoshioka, T. Tsuru, *Langmuir* 27 (2011) 13996. 8) T. Tsuru, T. Shibata, M. Kanezashi, T. Yoshioka, *J. Membr. Sci.* 421-422 (2012) 25-31 9) T. Tsuru, S. Sasaki, T. Kamada, T. Shintani, T. Ohara, H. Nagasawa, K. Nishida, M. Kanezashi, T. Yoshioka, *J. Membr. Sci.* submitted.

(142a) Nature-Inspired Geomimetic Separation Membranes

Tung, K. L., National Taiwan University

Originally presented on: 11/4/2013 14:30:00 - 15:00:00

There are many lessons that one could learn from, and/or mimic in nature. In the past decades, biomimetic concept has been widely adopted for biomaterials and polymeric materials designs by scientists and biologists. The biomimetic concept has promoted a fantastic progress in the science and engineering of bio/organic materials. Nevertheless, the geomimetic concept was astonishingly few to be taken into consideration for inorganic materials development. Recently, several new concepts to develop novel inorganic membrane materials have been proposed and new preparation methods from a fundamental understanding of formation mechanism and morphology of natural materials haven been designed. In this talk, several potential routes intentionally mimic existing geological processes to develop new inorganic membranes will be presented.

(155a) Charged Polymer Membranes for Water Purification and Power Generation

Paul, D. R., The University of Texas at Austin

Originally presented on: 11/4/2013 15:15:00 - 15:45:00

Providing sustainable supplies of water and energy are critical global challenges for the coming decades. Polymeric membranes are the dominant technology for desalination and could be useful for power generation applications as well. These desalination and power generation applications include reverse osmosis (RO), nanofiltration (NF), forward osmosis (FO), pressure-retarded osmosis (PRO), electrodialysis (ED), membrane capacitive deionization (CDI), and reverse electrodialysis (RED). Membranes with tailored water and salt transport properties, that are also chemically and physically robust, are required to extend and optimize these technologies. Most membranes for these applications will contain anionic or cationic charges covalently bound to the polymer. As will be discussed, the understanding of how fixed charges affect transport of ions is not currently adequate to guide the development of new materials and structures that will meet these future needs. Progress in this direction will be aided by use of physically realistic models combined with experimental membrane characterizations that go beyond measuring water flux and salt rejection. This presentation will elaborate on these issues and summarize results from ongoing investigations of structure-property relationships aimed at providing a better fundamental understanding of material structure and membrane performance.

(155b) Opportunities for Scalable Desalination

Belfort, G., Rensselaer Polytechnic Institute

Originally presented on: 11/4/2013 15:45:00 - 16:15:00

Reverse osmosis (RO) has matured from a laboratory curiosity at the University of Florida and the University of California at Los Angeles in the early 1960s to an accepted industrial desalination process widely considered as the least energy intensive and most economically attractive technology for separating salt from water. Plans are now in progress in Japan to build a megaton plant (1,000 m3/day). With this wonderful success story, where are the opportunities for improvement? They include development of smooth, low fouling, chorine stable and long lasting membranes. Can we replace interfacial polarization as a mean of producing salt rejecting rough and chlorine susceptible membranes? To address this question fully, we need to understand the limitations of RO, consider fundamental principles of the molecular behavior of water and salt at solid interfaces (i.e. synthetic membranes), and critically assess the limitations such as concentration polarization and fouling. In this presentation, we will first summarize the major technological breakthroughs associated with the success of large-scale RO including the reduction in energy use. While fouling and concentration polarization offer inherent limitations, they are dealt with through pretreatment (removal of foulants), chemistry and hydrodynamics. Despite its successes, the molecular basis of salt separation by RO is still controversial: Gluekauff, Matsuura and Sourirajan, and Belfort have suggested molecular mechanisms that depend on dielectric, bound water and ion hydration arguments, respectively. Given the importance of water-polymer interactions (hydrogen bonding), polar membrane materials such cellulose acetate and aromatic polyamides are widely used. However, the lack of a molecular-level understanding means that a rational design strategy for improving RO efficiency is missing. Recent experiments with oxidized graphene and single walled carbon nanotubes, together with simulations of water near polar and non-polar interfaces suggests exciting possibilities for high-flux desalination membranes. These results indicate that water moves through non-polar nano-pores with little or no friction, with very high water fluxes using small amounts of energy. Recent attempts to take advantage of these ideas are limited by concentration polarization. Increasing the mass transfer coefficient through improved fluid mechanics will be needed to remove the polarized salt and reduce salt leakage. Nature-inspired transport, through the well-known aquaporin channels in biological membranes, uses a mixture of intermolecular forces to optimize selectivity. In summary, we plan to discuss how integrating recent molecular insights with decades of industrial advances will provide exciting opportunities for scalable desalination.

(155c) Desalination Studies Using Membrane Distillation

Sirkar, K. K., New Jersey Institute of Technology
Singh, D., New Jersey Institute of Technology

Originally presented on: 11/4/2013 16:15:00 - 16:45:00

Membrane distillation for desalination employs heated brine in contact with a porous hydrophobic membrane to generate water vapor which is recovered in a variety of ways on the other side of the membrane. Direct contact membrane distillation (DCMD) employs colder distilled water on the other side to condense the water vapor. Air gap membrane distillation (AGMD) devices utilize a cold surface to condense the water vapor and generate the distillate. For brines at 50-93^{0}C, this presentation will focus primarily on hollow fiber membrane-based devices for DCMD as well as AGMD; studies in DCMD for these brine temperatures included synthetic brines, simulated sea water as well as actual sea water. For brines at 100-130^{0}C, this presentation will illustrate the DCMD performances of flat as well as hollow fiber membranes of PDVF and PTFE for a variety of salt concentrations, distillate in temperature, brine flow rate as well as simulated produced water from steam assisted gravity drainage (SAGD) process.

For AGMD the colder condensation surface is provided by a solid nonporous hollow fiber of polypropylene (PP). The evaporation fiber set consists of porous hydrophobic hollow fibers of either PP with a porous fluorosilicone coating or PVDF. Through the bores of the second set of solid nonporous hollow fibers flows the cooling fluid or the colder feed brine itself as it gets heated up. The performances of such two-hollow-fiber-based AGMD devices will be illustrated. The water vapor flux levels achieved in AGMD were quite high compared to those reported in literature. For the lower temperature DCMD studies, hot brine undergoes rectangular cross flow over the outer surface of highly porous hydrophobic PP hollow fibers whose outside surface was coated with porous plasma polymerized silicone-fluoropolymer coating to mitigate pore wetting and distillate contamination as cold distillate flows through the bores of fibers having a large wall thickness in DCMD. Extended pilot scale operation demonstrated no salt leakage, stable and repeatable performance with synthetic solutions having starting feed salt concentrations of up to 10% as well as with sea water concentrated up to 20% salt via batch recirculation. An overview of the device performance, resistance to scaling, concentration cascades and energy requirements will be provided for a variety of brines.

(142e) Novel Polymeric Hollow Fibers for Pervaporation and Multi-Bore Membranes for Membrane Distillation

Chung, T. S., National University of Singapore
Ong, Y. K., National University of Singapore
Wang, P., National University of Singapore
Originally presented on: 11/4/2013 16:45:00 - 17:15:00

Pervaporation membranes are often suffered severely due to solvent induced swelling because of the direct contact with the feed. We will review the progress of various polymeric membranes for pervaporation and then propose a novel single-step dual-layer co-extrusion technique called "immiscibility induced phase separation (I^2PS)" to fabricate high performance hollow fibers for the dehydration of ethanol. By taking the advantages of phase separation occurring in immiscible dopes made of incompatible polymers, dual-layer hollow fibers can be fabricated by simultaneously extruding immiscible blend dopes through a triple orifice spinneret. A selective layer is formed at the outer surface of the inner layer due to the incompatibility between both inner and outer-layers. As a result, the as-spun hollow fiber membranes consist of a protective layer on top of the selective layer which suppresses the solvent induced swelling during pervaporation. Hence, the flux of the hollow fiber can be enhanced by reducing the selective layer thickness without sacrificing the water/ethanol selectivity. Continuous performance tests demonstrate that the fibers spun from the I^2PS process possess stable dehydration performance throughout the monitored period.

Highly constrained by the requirements of high porosity and large pore sizes, traditional single-bore hollow fiber membranes often suffer from easy breakage and performance instability during long-term operations of membrane distillation. We will review the evolution of polymeric membranes for membrane distillation and then propose a new-generation PVDF multi-bore hollow fiber (MBF) membrane with a lotus root-like geometry via novel spinneret designs and optimal spinning conditions. The effects of various spinning parameters will be investigated on the membrane macro- and micro-structure, mechanical properties and direct contact membrane distillation (DCMD) and vacuum MD (VMD) performance. The MBF membranes have excellent mechanical rigidity and elasticity. Even for the MBF membrane with a thin wall of around 40 μm, the maximum load was as high as 2.4 N. Most importantly, the performance of DCMD of the MBF membrane was only slightly lower or even comparable to that of single-bore membranes. In addition, the MBF membranes exhibited superior stability in terms of vapor permeation flux and salt rejection during the continuous DCMD and VMD experiments with robust operational conditions.

We anticipate that these pioneering works may have profound implications to the development of hollow fiber membranes for pervaporation and membrane distillation.

(155d) Treatment of Produced Water and Frac Fluids By Ceramic Membranes for a Sustainable Water Reuse

Czermak, P., KMUB

Originally presented on: 11/4/2013 17:15:00 - 17:45:00

Oil and gas industries generate large amounts of wastewater as a byproduct in both onshore and offshore production operations. This kind of wastewater is commonly referred to as “produced water” (PW) or "fracfluid" (FF) in hydraulic fracturing processes. It is very difficult to treat and its characteristics changes by well to well. Treatment of PW or FF could improve the economic viability of oil and gas fields and lead to a new source of water for beneficial use. In general, produced water treatment is approached through deoiling and de-mineralizing before its disposal or utilization. Various technologies and methods exist for treatment of oil field produced water. Similar methods are used to treat fracfluids. Most oil removal technologies cannot achieve the separation required to meet water quality standards. The use of ceramic membranes for treatment of wastewaters is growing in certain applications and above all in those filtration processes where polymeric membranes cannot be applied.

The talk focuses on a) the efficient development of single and combined membrane treatment processes for PW and FF using tubular ceramic membranes (MF/UF/NF) and b) real time measurement of oil content in treated/untreated produced water streams using a new generation of oil-in-water monitoring system based on the light scattering technique.

(226a) New Developments on Membranes for CO2 Separation and Capture

Chen, Y.,
Ramasubramanian, K., The Ohio State University
Severance, M. A., The Ohio State University
Tong, Z.,
Vakharia, V., The Ohio State University
Zhao, L., The Ohio State University
Zhao, Y., The Ohio State University
Dutta, P. K., The Ohio State University
Ho, W. S. W., The Ohio State University
Originally presented on: 11/5/2013 8:30:00 - 9:00:00

This presentation covers new developments on carbon dioxide-selective membranes for: (1) hydrogen purification for fuel cells and (2) carbon capture from flue gas. We have synthesized CO_2-selective membranes by incorporating amino groups into polymer networks. The membranes have shown high CO_2 permeability and selectivity vs. hydrogen, carbon monoxide and nitrogen up to 180°C. We have elucidated the effect of amine steric hindrance in the solid membrane, showing significant enhancement for CO_2 transport. Hydrogen sulfide permeates through the membrane much faster than CO_2, allowing H_2S removal in the treated synthesis gas before water-gas-shift (WGS) reaction. Our initial experiments have shown a nearly complete removal of H_2S from 50 ppm in synthesis gas to about 10 ppb in the hydrogen product. Using the membrane, we have obtained <10 ppm carbon monoxide in the hydrogen product in WGS membrane reactor experiments via CO_2 removal. The data have been in good agreement with our modeling predictions. The membrane is being scaled up for commercialization. For CO_2 capture from flue gas in coal-fired power plants for sequestration, we will discuss a new concept of inorganic/polymer composite membranes along with initial membrane transport results. We will also discuss the process modeling along with cost sensitivity for the CO_2 capture to achieve the US DOE target of £35% increase on the cost of electricity (COE) and £$40/tonne CO_2 captured for 2025. Cost calculations show that this target can be met with an air-sweep process to achieve 90% CO_2 recovery and 95% purity of the CO_2 product. This process operates at slightly above atmospheric feed pressure (about 1.1 atm) with feed compression for cost optimization and requires a low membrane module price of \$2.5/ft^2 (\$27/m^2), a CO_2/N_2 selectivity of about 150 combined with a high CO_2 permeance of 3000 GPU.

(226b) Organic Solvent Nanofiltration Molecular Separations In Organic Liquids

Livingston, A. G., Imperial College London

Originally presented on: 11/5/2013 9:00:00 - 9:30:00

Organic liquids are ubiquitous in chemical science based industries, which range in scale from refining to pharmaceutical production. It is generally accepted that 40-70% of capital and operating costs in these industries are dedicated to separations; and a substantial fraction of this cost is related to processing of organic liquids, both as product streams and solvents. Organic Solvent Nanofiltration (OSN) technology has the potential to provide game changing alternatives to conventional concentration and purification technologies such as distillation, liquid extraction, adsorption and chromatography, reducing energy requirements and so improving sustainability.

In order to achieve this potential, membrane modules must meet several challenges. They must be stable in organic solvents, offer attractive fluxes and rejections for systems of interest, and give reliable and predictable service lifetime and performance. The obvious benefits of membrane processing have been apparent for many years, and have attracted research, development and commercialisation efforts from academic groups, end users, and membrane suppliers. In the last few years these efforts have resulted in a growing range of commercially available membranes, and an increasing number of industrial applications.

This presentation will firstly briefly describe recent developments leading to a new class of OSN membranes and modules. The DuraMem^{TM} series OSN membranes are made through post-formation processing including crosslinking of polymer membranes, which makes them stable in polar solvents such as acetone, DMF and THF. These membranes are commercially available from Evonik MET as spiral elements. Some of the key challenges overcome in developing a membrane from lab bench discovery to a commercially available product will be described.

The presentation will then describe recent research in OSN membrane formation, and OSN applications. New membranes include thin film composites, based on interfacial polymerisation (TFC-IP), having good flux and rejection in solvents including DMF and THF. There are a growing number of applications of OSN to industrial separation problems, and some will be outlined, together with the advantages OSN brings to commercial systems. Prediction of process performance for organic liquid systems will be discussed. Finally, the attributes of "ideal" membranes, and likely limitations on system performance will be outlined.

(226c) Sustainability Issues in Membrane Systems for Bioprocessing
Zydney, A. L., The Pennsylvania State University
Originally presented on: 11/5/2013 9:30:00 - 10:00:00

Although product yield and purity tend to be the dominant issues in the development of membrane systems for bioprocessing applications, membranes can also play a major role in the development of more sustainable bioprocesses. This talk will highlight several examples of sustainability issues and opportunities in membrane technology for bioprocessing applications. Ultrafiltration is currently used in the final concentration and formulation of nearly all recombinant protein products. In addition to giving very high product yield and quality, ultrafiltration also provides significant energy savings relative to alternative (e.g., thermal) processes for product concentration. Membrane chromatography can be used to significantly reduce the energy costs compared to traditional packed columns due to the significant reduction in the required pressure drop in the highly porous membranes. There are also exciting new opportunities for using membrane systems to develop continuous bioprocesses with dramatically smaller environmental footprints. These continuous processes are based on perfusion bioreactors, e.g., hollow fiber membrane bioreactors with the cells immobilized in the shell region of a hollow fiber module or more traditional bioreactors in which the cells are continuously recycled by a microfiltration unit. New research on continuous countercurrent tangential chromatography has demonstrated that this technology can provide continuous product capture using high performance resins that are processed as a flowing slurry through a series of static mixers and hollow fiber membrane modules. These continuous systems have productivities that are 10-100 times greater than conventional batch processes, dramatically reducing capital costs and environmental impact. Membrane systems should continue to play a critical role in the development of sustainable bioprocesses.

(226d) Membrane Separations As "Lever" Points in Biomass-Based Processes

Pellegrino, J., University of Colorado

Originally presented on: 11/5/2013 10:00:00 - 10:30:00

From a technical perspective, sustainability is a second law of thermodynamics analysis that has become synonymous with development of non-fossil fuel energy carriers. In that broad milieu, harvesting photons into biomass and converting these into different chemical forms has been a "grand challenge" for many decades. The challenge is having a "process make sense" from both an economic and life-cycle (environmental) perspective (the latter being where the second law usually shows up). I will illustrate how membrane-based separations can be "enabling" for potentially sustainable processes within both the algal and lignocellulosic biomass to fuels and chemicals framework. Useful membrane materials already exist for some applications, while optimum material properties and module designs are needed for others. Examples of techno-economic analysis of algal biomass processes will illustrate how membrane separations became a "lever" point for a different idea (for the appropriate species).

(155e) Progresses in Membrane Integrated Systems for Water, Energy and Raw Materials Production

Drioli, E., Institute on Membrane Technology of the National Research Council - (ITM-CNR)
Chung, T. S., National University of Singapore
Macedonio, F., Institute on Membrane Technology of the National Research Council - (ITM-CNR)
Originally presented on: 11/5/2013 10:30:00 - 11:00:00

In Nature complex molecular separation processes are controlling strategic aspects of life and evolution. Multifunctional membrane operations are often the core of the systems. Science and Engineering have been realized important progresses in the understanding of these phenomena and in their design for reproducing their potentialities artificially at industrial level or in medicine.

Artificial polymeric and inorganic membranes have been already studied and realised with important impacts on strategic aspects of our life. Dominant role has today membrane engineering for solving water stress by desalination and water reuse. Significant contribution to new form of energy systems might be related in the next future to the utilization of salinity gradient, via PRO or RED. The design of integrated membrane operations where fresh water production, energy production, and metals and various salts recovery will be realized simultaneously is becoming an interesting possibility.

New membrane reactors and membrane contactors will contribute to the realization of these objectives.

(299a) Solar- Powered Reverse Osmosis Technology for Brackish Water : The State of The Art

Abkar, L., Institute for Energy and the Environment New Mexico State University, Las Cruces, NM, USA.
Mallick, K.,
Alshukri, R.,
Ghassemi, A., New Mexico State University, Institute for Energy and the Environment/WERC
Loya, J., New Mexico State University,Institute for Energy and the Environment/WERC
Originally presented on: 11/5/2013 12:30:00 - 12:45:00

Water and energy are two of the most important factors for the development of any country. With increase in both population and industrialization, it's a challenge to produce ample freshwater with minimum energy cost. There are many places in the world that are turning into arid area due to water scarcity. Reverse osmosis desalination technology seem to be the best possible solution to this problem. With fossil fuel sources exhausting, renewable energy is most suitable to power RO for remote off-grid area. Solar powered– RO technology is applicable to area with ample solar resources. Many researches have been done for years to improve the PVRO system. Over the years this system has gone through many changes. For instance, batteries were taken down and direct connections were made between the PV panels and the pressure variable pump, turbines or hydraulic pumps were installed for energy recovery, etc. In this paper, a comprehensive review has been done on PVRO systems, its developments and water production cost in recent years.

(299b) Nanofiltration: High Potential, Poorly Defined and Understood
Paulson, D., Water Think Tank, LLC
Originally presented on: 11/5/2013 12:45:00 - 13:00:00

Nanofiltration: High Potential, Poorly Defined and Understood

The membrane mediated process known as nanofiltration is receiving much attention, but some of it due to the erroneous assumption that the technology came out of recent nano-materials advances, or that it performs based on the characteristics of nanoparticles. In reality, nanofiltration (NF) has been employed commercially for four decades, predating today's popular concept of nanotechnology. Even within the maturing membrane industry, nanofiltration is defined in varying manners. For an important technology to advance to full potential, such definitional incoherence is counterproductive.

The potential of nanofiltration is very high, and could be applied in more unique ways than its sibling membranes processes. To aid the research and industrial communities' understanding, this paper defines the current reality of the nanofiltration process, and addresses its history and its probable future. Few realize when the first NF membranes were developed, and for which applications and the origin of the term nanofiltration. Interestingly, the earliest NF membranes were involved in a patent suit that led to a seminal US Supreme Court decision, the only involving patent interpretation in a three decades span. These and other historical highlights are presented to help put the importance of this process in perspective.

The NF process is briefly described in relationship to the better understood membrane classes, reverse osmosis and ultrafiltration (RO and UF). Further definition is provided by description and comparison of NF's major applications in water treatment, industrial manufacturing and the more exotic emerging uses for medicines, fine chemicals and biofuels.

The lack of communication and common among the industries that use NF explains its incoherent definition. Lack of standardization and poor cooperation among standards setting organizations are described. Specific examples of misunderstandings that serve as roadblocks to the use of this process are given, including differences in definitions by standards groups, a membrane vendors and user groups (industries) are presented. Different views between the municipal water treatment community, industrial engineers, and the research community are described. A more universal definition, of both the process and its mediating membrane is proposed.

Whether defined as "nanotechnology" or not, separation of materials at the molecular level is a key to emerging technologies in several arenas of use. The paper concludes with a summary of recent advances, failed efforts, technical avenues to improve the membranes and the author's predictions for near and long-term improvements.

(299c) RO/NF Applications in Brackish Groundwater Desalination: Membrane Characterization

Vaseghi, G., New Mexico State University
Ghassemi, A., New Mexico State University, Institute for Energy and the Environment/WERC
Loya, J., New Mexico State University,Institute for Energy and the Environment/WERC
Originally presented on: 11/5/2013 13:00:00 - 13:15:00

Millions of people in the world suffer from the lack of potable and safe water. Water problem is practically the most important global issue nowadays, mainly because of decreasing fresh water resources and continuous deterioration of the quality of available resources. Water sources include oceans, brackish waters, and wastewaters. In the most arid and semi-arid regions of the USA such as New Mexico, California, Arizona, and Colorado, brackish water is an important source of feed water, and its most common quality problems are caused by suspended solids and hardness. Since desalination technologies are expensive, water shortage and scarcity pose significant threats for developing countries. Consequently, there is high interest in reducing the costs of water desalination.

Processes of desalination, or removing salt from water, include thermal processes, such as Distillation, and membrane processes, such as Reverse Osmosis (RO) or Electrodialysis (ED). Reverse Osmosis (RO) is one of the technologies used for desalinating brackish and saline waters to provide drinking water. RO treatment plants use semipermeable membranes and pressure to separate salts from water. They typically use less energy than thermal distillation, leading to a reduction in overall desalination costs, and allowing new brackish groundwater desalination facilities to use RO technology much more economically than distillation. Hence, this study improves the benefits of using RO by reducing the cost, time, and energy spending to find the best approach for different conditions. It focuses on the characterization of different types of membranes used in RO systems by testing effects of Temperature, pH, Conductivity, and Recovery on product. The objectives are accomplished by utilizing pilot plant experiments. A huge database collected from a full pilot-scale system, located at the Brackish Groundwater National Desalination Research Facility (BGNDRF) in Alamogordo, and operated by New Mexico State University are analyzed.

(299d) Investigating the Effecitive Parameters in Optimizing Reverse Osmosis Technology for Water Treatment

Abkar, L., Institute for Energy and the Environment New Mexico State University, Las Cruces, NM, USA.
Ghassemi, A., New Mexico State University, Institute for Energy and the Environment/WERC
Loya, J., New Mexico State University,Institute for Energy and the Environment/WERC
Originally presented on: 11/5/2013 13:15:00 - 13:30:00

As population growth, water scarcity become the major issue of concern. There are different types of desalination technologies such as thermal which includes multi- stage flash (MSF) and Multiple effect distillation (MED), and membrane based technologies such as reverse osmosis (RO) , electrodialysis (ED), electrodialysis reversal (EDR), membrane distillation (MD) and forward osmosis (FO) to provide fresh water. Among these processes Reverse Osmosis has shown the highest capabilities commercially. In this regard, optimizing RO system become a significant concern in order to decrease water production cost and make it more affordable. The process optimization depends on parameters such as temperature, pressure, membrane characteristics and feed water constituents. This paper presents a comprehensive review and assessment on the factors which affect the performance of the inland reverse osmosis system. A review on the major advancement in energy issues, water recovery, membrane characteristics, concentrate disposal, RO cost is done.

(299e) Carbon Electrode Driven Electrodialysis

Barber, J., GE Water & Process Technogies
Yang, H.,

Originally presented on: 11/5/2013 13:30:00 - 13:45:00

Electrodialysis processes require the application of an electric field in order to drive the migration of ions and cause deionization. The conventional approach is to utilize metal electrodes (Stainless Steel for the cathode and noble metals for the anode) and perform water electrolysis at the surface of the electrodes to create the electric field. The byproducts of water electrolysis, i.e. acid, base, and hydrogen, oxygen and chlorine gas, can cause system complexity, extra cost and environmental, health and safety concerns. To address these issues GE has develop a new capacitive carbon electrode to be used in electrodialysis reversal stacks which mitigates all of these concerns. In this presentation a detailed look at the electrodes design and operational parameters will be discussed, with reference to specific case studies from a number of field sites in rural and isolated communities.

(299f) Impact of Silica in Water Treatment Technology

Naderi Behdani, F., New Mexico State University,Institute for Energy and the Environment/WERC
Ghassemi, A., New Mexico State University, Institute for Energy and the Environment/WERC
Loya, J., New Mexico State University,Institute for Energy and the Environment/WERC
Originally presented on: 11/5/2013 13:45:00 - 14:00:00

In many parts of the world the use of water in reverse osmosis (RO), cooling, boiler, and geothermal applications is limited by the poor solubility of silica. Silica, the second most abundant element, is found in all natural water supplies. Silica solution chemistry is very complex and difficult to predict. In RO feedwaters, silica can exist in three different forms as follows: Monomer silica or silicic acid, polymeric silica that results from polymerization of silicic acid, and filterable silica. Ca^{2+} and Mg^{2+} ions have a strong influence on the formation of filterable silicate. It has been proved that the polymeric silica is mainly responsible for the membrane scaling. The challenge is how to exceed the range of silica solubility to facilitate the use of RO systems in areas where the available feedwaters have high silica concentrations. Silica solubility is well known to be influenced by both pH-value and temperature. This paper is a review on silica fouling problems and the use of various products for extending silica solubility.

(299g) Selectivity Comparison for Two Cation Exchange Membranes in the Electrodialysis Process

Karimi, L., New Mexico State University,Institute for Energy and the Environment/WERC
Ghorbani, A.,
Ghassemi, A., New Mexico State University
Loya, J., New Mexico State University,Institute for Energy and the Environment/WERC
Originally presented on: 11/5/2013 14:00:00 - 14:15:00

Electrodialysis is a membrane based separation in which charged species can be removed due to applied electrical potential. Although electrodialysis has been applied in very diverse industrial fields such as brackish and sea water desalination, salt production by concentering sea water or using brine discharged from a reverse osmosis seawater desalination plant, heavy metal recovery from waste water, recovering reverse osmosis concentrate stream, and citric acid separation, its high capability in brackish water desalination has been demonstrated. It is more applicable than reverse osmosis process for removing the impurities from saline water with high amount of silica. Saline water with high content of silica causes scaling problem in RO membranes, while ED is robust against scaling problems occurred due to silica content.

Depending on saline water resources, its composition can be very different and the levels of some ions can be more than the others. Therefore, they should be critically removed in order to reach the standard drinking water levels. Focusing on preferential ion removal in the ED operation makes it more purposeful to reduce the cost. Preferential separation of ions can lead to remove specific ions at appropriate rate and control the process cost. ED process is affected by several parameters; however, the ion exchange membranes have the most significant role in selective removal of ions in ED process. In fact, ion exchange membranes have significant role in selective removal of ions in this process. Various ion exchange membranes with diverse characteristics have different preference toward removing the charged species.

In this study the different operation of cation exchange membranes for CR67 and coated CR67 with Polyethyleneimine (PEI) were investigated. Their behavior toward preferential removal of different cations was verified. The experiments were done using a lab-scale ED equipped to a complete data acquisition which gives a significant level of confidence that both membranes were examined at same conditions. The water samples were analyzed using ion chromatography, IC 5000. The studied ion exchange membranes showed different behavior at toward different charged species. Sodium ion removal was considered as a reference ion in this study.

(299h) Developing a Mechanistic Transport Model for Electrodialysis/Electrodialysis Reversal Process

Ghorbani, A., New Mexico State University
Karimi, L.,
Ghassemi, A., New Mexico State University, Institute for Energy and the Environment/WERC
Loya, J.,
Originally presented on: 11/5/2013 14:15:00 - 14:30:00

Developing a Mechanistic Transport Model for Electrodialysis/Electrodialysis Reversal Process

Azadeh Ghorbani *, Leila Karimi, Abbas Ghassemi, Jim Loya

Institute for Energy and the Environment New Mexico State University, Las Cruces, NM, USA.
***azadeh12@nmsu.edu**

There is no doubt that the fresh water demand will continue to increases due to the growth of the world's population and the increase of the worldwide demand on consumer goods. Technologies that were originally developed to desalinate water such as Reverse Osmosis (RO) and Electrodialysis/ Electrodialysis Reversal, ED/EDR, are widely applied to remove contaminants other than salt from freshwater supplies.

ED/EDR is an Electrical-driven membrane technology that is effective with salt removal in which ions are transferred through ion exchange membrane. These process would cost less compared to the RO process but no complete model has been presented yet. Therefore, there is a definite need for presenting a model which would include mechanistic transport model.

Modeling is a critical step representing the process and the construction of system of interest. One of the significant features of a model is to enable the user to predict the effect of changes to the system without imposing cost.

Transport models relate fluxes through the active layers to driving forces and provide mechanistic descriptions of how material travels through the membrane media.

Mechanistic transport models may be used to predict how a particular membrane will perform in a new process, or may help to develop new membrane characteristics. Development of a reliable membrane transport model for describing mass transport phenomena in ED/EDR is a highly desirable goal since not only does this allow the analysis of performance characteristics of a membrane, but can also assist in the designing new membranes.

In this study, the existing models of RO were reviewed and investigated in order to develop a novel mechanistic transport model for ED/EDR processes.

(299i) Flexible Desalination Systems for Variable Salinity Sources

Chapman, M., Bureau of Reclamation
Leitz, F., Bureau of Reclamation
Tiffenbach, A.,

Originally presented on: 11/5/2013 14:30:00 - 14:45:00

Many water sources are less desirable as a potable water source because they vary in quality over time, such as in tidal reaches on coastal rivers. In other situations water needs to be treated intermittently so that either water from several sources is treated at a common treatment location or equipment is moved to the source of contaminated water such as in centralized produced water treatment and small scale industrial waste treatment. Irrigation return flows vary with the amount of precipitation and irrigation and fertilization schedule. Over a longer scale of time, communities that use ground water may find the salinity increase due to seawater intrusion or declining aquifer levels. In all of these scenarios the treatment process and equipment needs to be adaptable to a wider range of source water qualities.The Variable Salinity Desalination (VSD) project was designed to evaluate modifications to the Expeditionary Unit Water Purifier- Generation 1-A (EUWP), designed to treat any source up to 60,000 mg/L total dissolved solids, 100 NTU turbidity, and temperatures, at between 45-60 percent recovery, to allow higher water recovery operation when treating brackish water. The system was modified with a bypass around the pressure exchanger to attain 75% water recovery to convert the system from a one stage design to two stages. Operability, ease of transition between designs, power requirements, and water quality were evaluated using a brackish groundwater feed source. Performance was compared to operation in the original configuration at three other sites with a wide range of salinity. Power use and water quality were as would be expected for the higher recovery. Options for building in design flexibility and cost impacts are discussed.

(299j) An Innovative, Environmentally Friendly, Low-Cost Desalination System

Nossoni, G., Manhattan college
Markes, K., Manhattan college
Matinez, H., Manhattan College
Piazza, A., Manhattan College
Qasim, M., Manhattan College
Abulencia, J. P., Manhattan College
Originally presented on: 11/5/2013 14:45:00 - 15:00:00

The lack of clean drinking water is rapidly becoming one of the most important international health issues. Many remote regions in developing countries face water shortages and lack of potable water causing significant hardship to local people. One popular method used is the desalination of brackish water. The most widely used and commercially proven desalination technologies consist of thermal (evaporative), membrane based methods such as electrodialysis (ED) and reverse osmosis (RO). Unfortunately, these processes require a large amount of energy, which translates to being expensive. People in rural areas of developing countries do not have the means or resources for expensive water treatment systems

To this end, a multi-stage *hybrid ED-CON* (Electrondialysis/Concrete) process was designed to separate salt from seawater. The typical setup of an electrodialysis (ED) cell is adopted for the proposed *ED/CON hybrid* cell. However, rather than using a series of anion and cation exchange membranes that are arranged in an alternating pattern between anodic and cathodic electrodes, two concrete slabs of prescribed thickness are used as permeable membranes separating the anodic, cathodic, and desalination chambers. When a DC potential is applied between the two electrodes, positively charged cations (Na^+) move toward the cathode passing thorough the cathodic concrete membrane, and negatively charged anions (Cl^-) move toward the anode passing through the anodic concrete membrane therefore, the concentration of salt decreases as the salt water flows through each stage in the series of desalination cells. This design is unique and environmentally friendly because it can be produced using locally available recycled material and can be powered by a renewable energy source such as a solar panel.

Concrete is a mixture of Portland cement, water, and aggregate. It can be produced locally from abundant natural or recycled materials. As the water-cement ratio in a concrete mixture increases, the porosity of the cement paste increases. Concrete pores are interconnected and dissolved ions can move through concrete with water if there is a pressure gradient, or by diffusion due to a concentration gradient when a pressure head is absent.

Extensive testing was performed to optimize the process step by step. The parameters chosen for study were concrete porosity, the type of anodic solution, and potential used in the test. These variables were adjusted as the research progressed. The suitability of the parameters was judged based on the conductivity of the solution in the desalination chamber, which was used as a surrogate measure for the salinity. Preliminary results were promising and they show that the conductivity of salt water can be reduced by 10 ms/cm in 6 hours, which corresponds to about a 27% reduction in the salinity of water, in an optimized system. The optimized system used pervious concrete slabs, lime water in the anodic chamber, and a 10 V potential between the electrodes. The idea of using concrete as a membrane in water filtration is novel and addresses the issue of sustainability. This concept can be broadened and applied not only to the desalination of water, but also to the removal of other contaminants in ground or surface water by designing the concrete filters with suitable aggregate material and size gradations.

(702a) Low-Pressure Sweep Membrane Modules for Carbon Dioxide Capture

Kniep, J., MTR
Hao, P., MTR
Chan, K., MTR
Fulton, D.,
Freeman, B., Electric Power Research Institute
Merkel, T., MTR
Baker, R., MTR
Originally presented on: 11/7/2013 12:30:00 - 12:55:00

To mitigate the harmful effects of global climate change, CO_2 in power plant flue gas must be captured and either sequestered or utilized in processes such as enhanced oil recovery. Membrane technology is an attractive CO_2 capture option because of advantages such as energy-efficient passive operation, tolerance to acid gases and oxygen, no hazardous chemical emissions or handling issues, minimal water requirements, and no steam use requiring modifications to the existing boiler.

Working with DOE, MTR has previously demonstrated the feasibility of a two-step membrane process design to recover CO_2 from power plant flue gas. The second step of this process is a selective sweep membrane module that strips CO_2 from the flue gas and recycles it back to the boiler with the incoming combustion air. MTR is developing a membrane module specifically designed for high gas flow, low pressure, sweep operation. These membrane modules contain up to 500 m^2 of membrane area, which is 10 to 20 times the membrane area of modules currently being used for flue gas CO_2 capture. These large area modules will reduce the manifolding complexity, footprint, and cost of the very large membrane plants required for flue gas treatment. By lowering the pressure drop through the modules, less compression/blower energy is needed to move gases through the module, which reduces the parasitic energy loss of the capture process. In this presentation, results of the development work to date and future plans, including slip stream field testing at the National Carbon Capture Center (NCCC) operated by Southern Company in Birmingham, AL, will be discussed.

The development of large, efficient, low cost, sweep membrane modules is an enabling technology for membrane CO_2 capture and sequestration systems. MTR is investigating several alternative process designs to remove CO_2 from coal power plant flue gas, and while all of the designs use a membrane contactor to recycle CO_2 to the power plant boiler, the second CO_2 separation unit need not be a membrane unit. The selection of different hybrid separation technologies, hybrid process designs, and process economics will be presented.

(702b) Major Interest of Hollow Fiber Membrane Contactors for An Ammonia-Based CO2 Capture Process: Detailed Analysis of An Innovative Process

Makhloufi, C.,
Roizard, D., Nancy Universite
Favre, E., Laboratoire Reaction et Genie des Procedes LRGP- CNRS, Universite de Lorraine
Originally presented on: 11/7/2013 12:55:00 - 13:20:00

Up to now, the reference technology for CO_2 capture in postcombustion makes use of monoethanolamine (MEA) as chemical solvent in a packed column. Despite high capture efficiencies, MEA induces too high energy penalties. Consequently, an abundant literature is discussing the use of aqueous ammonia for flue gas scrubbing. Indeed ammonia is cheaper than MEA, shows a higher absorption capacity and, most of all, its energy regeneration demand is much lower. Nevertheless, the high ammonia volatility is a considerable drawback which causes a high NH_3 slip within the flue gas. Specific operatic conditions are required to limit NH_3 losses and a post absorber scrubbing step is therefore needed.

Amongst growing technologies, membrane contactors appear as a promising technology able to allow significant process intensification when compared to conventional gas/liquid contactors. So far, no study dealing with the use of membrane contactors for the CO_2 capture in aqueous ammonia has been disclosed. For the first time, a detailed analysis of an ammonia based CO_2 capture process using hollow fiber membrane contactors will be given during this presentation. Particularly, membrane contactors ability to both improve CO_2 mass transfer performances and mitigate NH_3 volatilization compare to conventional gas/liquid contactors will be discussed.

First of all, it will be shown that CO_2 absorption in ammonia can not be performed using conventional microporous hollow fiber membrane contactors and composite membranes made of a thin polymeric layer facing the liquid phase must be used. In that sense, composite hollow fiber membrane contactors have been specifically designed, prepared and tested to favor CO_2 absorption and, in the same time, to limit the ammonia slip in the flue gas.

This involved precisely selecting the dense layer and it will be shown how an inadequate material selection could lead to its significant degradation. The chosen dense skin must be thin, highly permeable to CO_2 and selective for CO_2 over NH_3. This should allow reaching high CO_2 mass transfer coefficient while reducing drastically the ammonia slip in the flue gas. This was undoubtedly one of the main challenges. Indeed, it has to be recalled that a faster permeation of CO_2 compared to NH_3 in dense polymers was totally unexpected and to our knowledge unexplored. For the first time, 4 chemically resistant polymers showing a reverse selectivity CO_2/NH_3 were identified. These meaningful results obtained thanks to time lag experiments will be presented in this presentation.

In a second step, CO_2 absorption experiments in aqueous ammonia using lab scale composite hollow fiber membrane contactors (CHFMC) have been successfully carried out. The effect of the main operating conditions namely ammonia concentration, temperature, gas and liquid flow rates, nature of the dense layer, on the CO_2 absorption performances and the ammonia loss has been studied. The overall volumetric CO_2 absorption capacity in CHFMC has been compared to the one obtained in packed columns. It has been concluded that membrane contactors strongly enhance the CO_2 absorption performances in comparison to conventional contactor. Moreover, a sharp decrease of the ammonia slip has been observed. The above results will be detailed in the presentation and the performances and advantages of this new CO_2 absorption process will be discussed further.

(702c) Membrane Assisted Liquid Absorbent Regeneration (MALAR):Innovative Use of Membranes in Liquid Absorbent Based CO2 Capture

Feron, P.,
Zhao, S.,
Originally presented on: 11/7/2013 13:20:00 - 13:45:00

CO_2 capture and geological storage (CCS) is a necessary option to limit CO_2 emissions from existing power stations. In the CCS chain the capture and compression of CO_2 to pipeline or injection pressure represents around 70% of the costs and all of the energy requirements of the process. Typical output reductions for 90% capture of CO_2 at a power station are around 30% and these are very large indeed. The leading capture technology is based on a liquid absorbent which chemically reacts with CO_2. The liquid absorbent is regenerated thermally and it will be re-used during many cycles. The CO_2produced from this process is also of high purity and can easily be stored in the underground after compression.

The energy required for the liquid absorbent regeneration is extracted as steam from the power station. This steam is then not available for power generation. Hence the output of the power station is reduced, which is undesirable. It is therefore necessary to develop absorbent regeneration processes which have much lower energy requirement. The starting point is to analyse the thermal energy requirement in more detail. It has the following contributions:

- The energy required to break the bond between CO_2 and the active component in the solvent as determined by the chemical formulation of the liquid absorbent.
- The heat required for bringing the solvent up to the regeneration temperature, which is a function of the CO_2 loading of the liquid absorbent.
- The evaporation enthalpy for the stripping steam which leaves the stripper together with the CO_2, which is determined by the temperature of the steam-saturated CO_2-product leaving the stripping column.

The overall energy penalty also incorporates the electricity use of the capture and compression process. This penalty is however to a large extent determined by the heat required to release the CO_2 from the liquid absorbent, typically around 4 MJ/kg CO_2. Dependent on the fuel type, this represents between 20 and 40% of the energy contained in the incoming fuel.

The ongoing development of new liquid absorbents will very likely reduce the first two contributions, i.e.

new amines with lower binding energy and higher CO_2-loadings have been identified. However the energy required for the production of stripping steam is still needed. In fact all the energy needed for the regeneration is brought in via the stripping steam and is transferred to the liquid absorbent via condensation. Hence efficient control of the steam balance is important for the energy performance of the capture process.

The MALAR process builds on this foundation, aiming to reduce the energy requirement to less than 2 MJ/kg CO_2 using a **M**embrane **A**ssisted **L**iquid **A**bsorbent **R**egeneration process. This process concept lends itself for application with any type of liquid absorbent.

The proposed innovation aims to efficiently integrate heat recovery in the liquid absorbent regeneration process. This is achieved by the use of membranes for transfer of heat using condensation and evaporation processes. The process concept is an entirely new one utilising membranes as gas-liquid contactors in such a way that the (common) occurrence of liquid leakage through the membrane is not an issue.

In the MALAR process energy is recovered via transfer of steam through a membrane. In the membrane condenser latent heat is recovered from the wet CO_2 product into a portion of the cold CO_2-rich liquid absorbent; in the membrane evaporator latent heat is recovered from the hot CO_2-lean absorbent into the stripping column. The use of membranes for direct heat transfer via evaporation or condensation enables quite small approach temperatures which would be unachievable and/or uneconomical with conventional heat exchangers. The compactness of the membrane equipment further adds to the achievement of small temperature differences and thereby a very energy efficient process. Two porous membrane configurations using either hyrodophobic or hydrophilic are proposed for the exchange of latent heat.

This application of membranes is a so-called membrane contactor application, i.e. the membrane is used to contact a gas flow with a liquid flow. Such operations often suffer from minor amounts of liquid permeating through the membrane to the gas phase. This liquid is subsequently carried away by the gas flow, which is generally not desired, and particularly if the gas enters the atmosphere. It will furthermore hamper the efficient transfer of components between the two phases. In the MALAR application envisaged here, liquid leakage is not critical as the heat transfer function is not affected and the liquid will be contained in the stripper column and mixed with the liquid absorbent already present.

A 50% reduction in thermal energy required for the liquid absorbent regeneration can be achieved by using the MALAR concept in combination with improved liquid absorbents, resulting in a thermal energy requirement below 2 MJ/kg CO_2. The improvements are coming from:

- A lower gas temperature exiting the stripper column reducing the evaporation loss
- A closer temperature approach through the use of the membrane heat exchangers
- The liquid absorbent having a double CO_2-loading and lower reaction enthalpy, which a feature of new liquid absorbents developed over the last 2 decades.

In summary: the benefits of the MALAR are:

- Energy efficient capture process through novel heat recovery method
- Smaller size equipment through use of high specific surface area membrane equipment
- Applicable to a wide range of liquid absorbents
- Different membrane types (hydrophobic and hydrophilic) are feasible
- Use of a membrane contactor in a non-critical application with respect to liquid leakage

(702d) Pressure Swing Membrane Absorption for Pre-Combustion CO2 Capture From Shifted Syngas

Sirkar, K. K., New Jersey Institute of Technology
Chau, J., New Jersey Institute of Technology
Jie, X., New Jersey Institute of Technology
Obuskovic, G., New Jersey Institute of Technology
Originally presented on: 11/7/2013 13:45:00 - 14:10:00

An advanced pressure swing membrane absorption (PSMAB) technology utilizing a hollow fiber membrane contactor device and a cyclic absorption-desorption process is being developed to produce purified hydrogen at a high pressure for IGCC-CCS plant's combustion turbine from low temperature (L-T) post-shift reactor synthesis gas and simultaneously obtain a purified CO_2 stream suitable for sequestration. The low temperature water gas shift reactor product stream is likely to be available at ~20 atm and around 150-200^{0}C. The experimental studies are employing He as a surrogate for H_2. Studies have been carried out with a 40%CO_2-60% He feed gas mixture coming in at temperatures up to 100^{0}C and pressures varying from 100 to 250 psig (689-1723 kPag). The nonvolatile absorbent liquid consists of an ionic liquid 1-butyl-3-methylimidazolium dicyanamide ([bmim][DCA]) (among others) containing nonvolatile amines. Two different cyclic processes have been studied; most of the performance will focus on a 5-valve cycle. Process performance studies have focused on the duration of various parts of the cycle, the effects of pressure and temperature variation, and the effect of hollow fiber membrane module design on the quality of the two product streams namely, the He-rich stream and the CO_2-rich stream. A numerical model has been developed to predict the process performance of the simpler 3-valve cycle for pure ionic liquid as the absorbent. Solubility and diffusion coefficient data for the gases are needed in this model; these were determined for the pure ionic liquid absorbent using a pressure-decay-dual transducer apparatus at temperatures of 323, 353, 363, and 373 K and at pressures up to 1.38 MPa (~200 psig).

(702e) Application of Composite Polymeric Membranes for Efficient Regeneration of Physical Solvents

Pesaran, A.,
Tande, B., University of North Dakota
Originally presented on: 11/7/2013 14:10:00 - 14:35:00

The most dominant CO_2 capture technology used for pre-combustion capture involves the application of physical solvents. Unlike chemical solvents, physical solvents do not react with the solute and they physically dissolve the acid gases, which are then stripped without the need to be heated by means of pressure swing techniques or combination of heat and pressure letdown. However, despite low energy required to regenerate the physical solvents and their high capacity to capture and separate acid gases from the syngas produced in a gasification plant, physical solvents have some disadvantages including CO_2 pressure loss and the energy required to pump the solvent to the high pressure absorber.

The primary objective of this work is to evaluate the use of composite polymeric membranes, for the recovery of CO_2 from CO_2-rich solvent streams. To achieve this purpose, an experimental bench-scale setup was built to investigate and quantify CO_2 removal capacity from the rich solvent across different types of membranes. This setup primarily consists of a high pressure absorber where solvent is loaded and saturated with CO_2 and the membrane module where CO_2 diffuses through the membrane. In order to increase the driving force across the membrane, sweep gas is used (Nitrogen). To evaluate the absorption process, absorber temperature and pressure is measured and recorded continuously using a NI data logger and the LabView software. The absorber is equipped with a home-made cooling water coil and a relief valve to adjust the absorber temperature and pressure respectively. Two 400 W inline pencil heaters are used and controlled with a temperature controller to adjust the solvent temperature downstream of the membrane module. The solvent line temperature and pressure are also recorded continuously. CO_2 concentration in the sweep gas stream is measured using the Non-Dispersive Infrared CO_2 analyzer and Agilent 7850A GC- FID depending on the concentration of CO_2 in the sweep gas. Dimethyl ether of polyethylene glycol (Selexol) is used as the solvent since it is reputed to be one of the major physical solvents for CO_2 removal. To evaluate the effectiveness of different types of membranes, CO_2 permeation rate and membrane selectivity were measured for different membranes. To better understand the effect of system pressure and solvent flow rate on different experiment responses such as CO_2 flux, selectivity and % recovery, a two-factor two -level full factorial design with two replicates an three center points were performed with the membrane that showed the most promising results in the screening study.

The results of the screening study indicated that PDMS based membranes (PERVATECH and PERVAP

4060) shows higher CO_2 permeability compared to PVAOH based membranes (PERVAP 1211 and PERVAP 1201). The best membrane for further analysis and experiments to find the optimum operational conditions was chosen as PEVAP 4060 from SULZER due to its high CO_2 flux and selectivity compared to other membranes.

A statistical analysis was performed to identify the significant factors for each individual response such as permeation rate, leak rate and selectivity. For CO_2 flux, pressure appeared to be strongly significant. However, solvent flow rate had no significant effect on the rate of CO_2 permeation. The immediate conclusion from this observation is that the main mass transfer is controlled by the membrane. With respect to the solvent leak, the analysis of Pareto charts suggested pressure to be significant and solvent flow rate to be insignificant. As the system pressure increases, the liquid on the upper chamber of the membrane module forces itself more into the membrane and hence, solvent leak increases. However, increasing the solvent flow rate leads to higher liquid velocity on top of the membrane and shows no significant changes on the rate of solvent leak. Neither system pressure nor solvent flow rate found to be significant considering the selectivity as the experiment's response. Finally, regarding the percent recovery, both the system pressure and solvent flow rate appeared to be significant. This is mainly because at elevated pressures, mole fraction of CO_2 in solvent increases and thus a higher driving force for CO_2 permeation exists. As a result of this, percent of recovery increases by pressure as confirmed by the Main Effects plots. On the other hand, by increasing the solvent flow rate, more amount of CO_2 enters the upper membrane chamber but the mass transfer resistant through the membrane prevents more CO_2 from being transported. Thus, there is the condition of introducing more CO_2 to the upper chamber but still a slow mass transport process being controlled by the membrane which eventually decreases the separation ability of the membrane (percent recovery).

In order to examine the chemical stability and structural integrity of the membranes after being exposed to the high pressure solvent, a series of post-experiment characteristic tests such as FTIR and DSC were performed. The results of these studies revealed no major changes.

(702f) Efficient Capture of CO2 From Pulverized Coal Fired Plants

Dada, E. A., ChemProcess Technologies (CPT), LLC
Olanrewaju, K., University of Iowa
Drayton, R. A., THS, Inc
Originally presented on: 11/7/2013 14:35:00 - 15:00:00

As demand for fossil fuel based energy increases, the need to sustainably manage and mitigate carbon dioxide (CO_2) emissions, the largest volume contributor to global warming, is unequivocally necessary. Various studies discussing different methods of capturing/sequestrating $CO_{2,}$ either by separation from a stream of effluent flue gases after a combustion process or through fixation, have been thoroughly reviewed. These methods range from chemical and physical absorption and adsorption, to forestation, Ocean fertilization, photosynthesis process, mineral carbonation and other *micro*-scale sequestration schemes. As this study will focus more on the technical options of separating CO_2 from pulverized coal-fired plant (a *post* combustion process), the current method of chemically absorbing CO_2 from stream of flue gases is fraught with challenges such as: the energy intensity of the process, limited loading capacity, solvent degradation and the corrosive effect of the solvent upon the process units. While physical absorption and adsorption address some of the limitations within chemical sorption process, the shortcomings still persist in their inability to handle large volume of flue gases with low CO_2 partial pressures. This work discusses two novel solutions that address the issues concerning the current technical approaches of sorbing carbon from effluent flue gases at coal-fired plants. It is proposed to design an energy efficient process that is cost effective and novel solvent formulation that an enhance CO_2 loading capacity due to their higher chemical stability and lower corrosive characteristics.

(780a) Continuous Enzymatic Hydrolysis of Biomass in a Membrane-Assisted Reactor

Malmali, M., University of Arkansas
Stickel, J., National Renewable Energy Laboratory
Wickramasinghe, S. R., University of Arkansas
Originally presented on: 11/8/2013 8:30:00 - 8:55:00

Although enzymatic hydrolysis of lignocellulosic biomass is usually conducted in batch mode, continuous operation is desirable for a number of reasons. Higher overall investment costs due to larger reactor volumes, higher running costs due to frequent startup/shut down, one-time use of enzymes as well as enzyme separation costs are disadvantages of batch operation. Recovery and reuse of the enzymes during enzymatic hydrolysis of biomass is attractive as the enzyme cost is a major economic constraint. In addition, high sugar concentrations are desirable for subsequent fermentation steps, but obtaining high sugar concentrations directly from batch enzymatic hydrolysis is limited by product inhibition of the enzymes and material handling of slurries at high-solids concentrations. Development of a continuous saccharification process with integrated membrane separations overcomes the limitations of batch operation. Membrane separation steps integrated with the enzymatic hydrolysis reaction could enable recovery and reuse of enzymes, as well as concentration of product sugars. Further capital cost will be lower due to smaller process stream volumes as a result of higher concentrations.

In this study we have designed and tested an integrated continuous membrane reactor. The integrated membrane reactor includes a microfiltration/ultrafiltration membrane for recovery and recycle of the enzyme as well as a nanofiltration membrane for concentration of the sugars. Here, we present results for our continuous integrated membrane reactor. These results are compared to a batch reactor. We show that glucose production rates and product sugar concentrations are higher for the integrated continuous membrane reactor.

(780b) Small-Scale Production of Ammonia From a Combined Reactor-Membrane-Adsorber

Himstedt, H., University of Minnesota
McCormick, A. V., University of Minnesota
Schmidt, L. D., University of Minnesota
Cussler, E. L., University of Minnesota
Originally presented on: 11/8/2013 8:55:00 - 9:20:00

Ammonia is one of the world's most important chemicals; over 130 million metric tons are produced annually. Without modern fertilizers derived from ammonia two billion people would starve. Ammonia is currently produced via the Haber-Bosch process in large capital-intensive factories. Despite a century of optimization, single-pass ammonia production is only roughly 20% and requires large amounts of sacrificial fossil fuels to obtain raw materials.

We present work on a hybrid membrane/adsorption reactor system capable of producing ammonia from wind power. Hydrogen and nitrogen are obtained from electrolysis of water and membrane separation of air, respectively. The ammonia produced can be separated by either a selective membrane or pressure-swing adsorption. We believe membrane separation is more effective for larger scale operation, but the simpler absorption is better for smaller scale. By combining reaction and absorption into a single unit operation, ammonia conversion can be increased. The system performs as a mixed-media membrane reactor without the potential physical limitations of a membrane structure. Such a system may make sustainable distributed production of ammonia possible.

(780d) Membrane Processes for Conversion of Solar Energy to Electricity

Lipscomb, G., University of Toledo
Patil, R., University of Toledo
Originally presented on: 11/8/2013 9:45:00 - 10:10:00

The world faces increasing demand for energy as the population grows and becomes more affluent. This demand poses significant challenges as the consumption of fossil fuel energy sources ultimately is not sustainable and has the potential to effect climate change.

The conversion of solar energy to electricity offers one alternative to fossil fuel use. Currently, solar energy is converted to electricity using both thermal and photovoltaic processes. We propose an alternative route that uses membrane processes.

The proposed process utilizes a combination of either reverse electro-dialysis or pressure retarded osmosis to mix two working fluids of different salinity to produce electricity. The salinity difference is restored by demixing using either membrane distillation or membrane crystallization. This demixing is driven by solar thermal energy. The overall process leads to a conversion of solar energy to electricity.

The capabilities of the proposed process are explored through simulation. The relationships between membrane transport properties, area requirements, and working fluid properties are determined from which electricity generation costs can be estimated.

(780e) Industrial Membrane Hybrid Driers

Campos, D.,
Nemser, S., Compact Membrane Systems, Inc
Pennisi, K. J.,
Majumdar, S., Compact Membrane Systems, Inc
Originally presented on: 11/8/2013 10:10:00 - 10:35:00

This paper describes a chemically and thermally resistant membrane system for the dehydration of organic solvents and chemicals. The hydrated chemical can be fed to the system either in the liquid or the vapor phase. The feed to the membrane is separated into an organic rich fraction in the retentate and a water rich fraction in the permeate. The solvent recovery depends on the permeability of the membrane to the solvent and the purity required in the solvent-rich fraction. Applications of this technology include (but are not limited to) drying organics that are difficult to dry by conventional technologies (e.g., azeotropes or solvents that have boiling points close to water), heat sensitive chemicals, reacting mixtures in which the chemical reaction is inhibited by the presence and/or generation of water. This technology is been used commercially for the successful dehydration of lubricating oils. Pilot scale test data are presented on the application of this technology for drying ethanol, IPA, THF, toluene and phosphate esters. The membrane system can be combined with conventional distillation to create a hybrid process for removing water from organics resulting in cost and energy savings. Examples of this hybrid drying process are presented along with the economic advantages derived from using the hybrid technology.

(780f) A Generic Approach of Hybrid Membrane / Condensation Process : Potentialities and Limitations

Belaissaoui, B., Laboratoire Reaction et Genie des Procedes LRGP- CNRS, Universite de Lorraine
Favre, E., Laboratoire Reaction et Genie des Procedes LRGP- CNRS, Universite de Lorraine
Originally presented on: 11/8/2013 10:35:00 - 11:00:00

In numerous industrial applications, gaseous streams containing an inert gas and a condensable compound have to be treated by a separation process. A direct condensation strategy is often applied for that purpose, but it requires in certain cases a cryogenic operation (e.g. a liquid nitrogen cold box), typically when the target condensor temperature is lower than –30 C. This situation occurs for low boiling compounds and/or when a low residual concentration of the condensable compound is necessary. In that event, an alternative approach based on hybrid membrane condensation process can be proposed: the principle consists to achieve a pre-concentration of the condensable compound by membrane separation (vapor permeation), in order to enable condensation to occur on the permeate side of the membrane at a higher temperature. This concept has been already succesfully applied for Volatile Organic Compounds (VOC's) recovery from air [1-4], where increasingly stringent regulatory requirements make recovery of VOC's of utmost importance. Nevertheless, the hybrid membrane condensation concept could possibly be of interest for other industrial applications such as drying of natural gas, hydrocarbon removal from hydrogen purification, ammonia recovery or carbon post combustion capture [5].

In this study, a generic approach of a hybrid membrane/condensation process has been developped and will be compared to a direct condensation strategy, taken as the baseline. It will be shown that, based on only two key parameters, namely the condensable compound solubility parameter and the membrane selectivity, the potential interest of the hybrid approach can be rapidly evaluated. The Hildebrand solubility parameter includes indeed the cohesive energy of the condensable compound and its molar volume; this property can be related to the condensation temperature, the heat of condensation or the solubility of a compound in an elastomeric membrane.

A large range of solubility parameter data, covering gases (CO_2, NH_3), VOC's (hexane, dicholoroethane, acetone or vinyle chloride) and water has been investigated through a series of simulations of the hybrid process. The influence of different membrane process parameters, e.g., recovery ratio, pressure ratio, stage cut has been carried out. The condensation temperature has been evaluated for each set of conditions and each compound, and, from this, the corresponding overall energy requirement of the process is estimated.

As a major outcome of this study, a generic process selection map will be presented. Based on the solubility parameter and membrane selectivity, the potentiality of the hybrid membrane/condensation process to achieve attractive performances comparing to the standalone condensation process is analysed. Heuristic rules and recommandations leading to an increased energy efficiency of the hybrid membrane condensation process covering diffrents solvent/inert/membrane systems will finally be proposed.

References

[1] Richard W. Baker, Noriaki Yoshioka, Judith M. Mohr, Amy J. Khan, Separation of organic vapors from air, Journal of Membrane Science, Volume 31, Issues 2–3, May 1987, Pages 259-271.

[2] H. Paul, C. Philipsen, F.J. Gerner, H. Strathmann, Removal of organic vapors from air by selective membrane permeation, Journal of Membrane Science, Volume 36, 1988, Pages 363-372.

[3] K. Kimmerle, C.M. Bell, W. Gudernatsch, H. Chmiel, Solvent recovery from air, Journal of Membrane Science, Volume 36, 1988, Pages 477-488.

[4] M. Leemann, G. Eigenberger, H. Strathmann, Vapour permeation for the recovery of organic solvents from waste air streams: separation capacities and process optimization, Journal of Membrane Science, Volume 113, Issue 2, 15 May 1996, Pages 313-322.

[5] B. Belaissaoui, D. Willson, E. Favre, Membrane gas separations and post-combustion carbon dioxide capture: Parametric sensitivity and process integration strategies, Chemical Engineering Journal 211–212 (2012) 122–132.

(784a) Characterizing the Gas Transport and Mechanical Properties of Thermally Rearranged (TR) Polymers

Liu, Q., The University of Texas at Austin
Paul, D. R., The University of Texas at Austin
Freeman, B. D., The University of Texas at Austin
Originally presented on: 11/8/2013 8:30:00 - 8:55:00

Polyimides containing ortho-positional functional groups, such as hydroxyl and acetate groups, undergo thermal rearrangement to form polybenzoxazole structures upon heating between 350 and 450°C. These thermally rearranged (TR) polymers show excellent gas transport properties and have strong resistance to plasticization. In this study, a series of TR homopolymer and copolymer were prepared with polyimide precursors based on HAB (3,3'-dihydroxy-4,4'-diaminobiphenyl), APAF (2,2'-bis (3-amino-4-hydroxyphenyl)-hexafluoropropane), and 6FDA (2,2'-bis-(3,4-dicarboxy-phenyl) hexafluoropropane dianhydride). Pure gas permeabilities of H_2, N_2, O_2, CH_4, and CO_2 were measured over pressure ranging from 3 atm to 17 atm at 35°C. Upon thermal rearrangement, APAF-6FDA has the best gas transport properties, presumably due to an enlarged cavity size and distribution of the TR polybenxazole structure. In addition, this study systematically documents the mechanical properties of the polyimides and TR polymers. HAB-6FDA polyimide has 10% elongation-at-break, whereas APAF-6FDA polyimide has only 4% elongation-at-break.

(784b) Vapor Permeation of Ethanol and Water in Thermally Rearranged (TR) Polymers

Gleason, K. L., The University of Texas at Austin
Paul, D. R., The University of Texas at Austin
Freeman, B. D., The University of Texas at Austin
Originally presented on: 11/8/2013 8:55:00 - 9:20:00

In this work, we present a high temperature permeation study of water and ethanol vapor in a thermally rearranged polyimide synthesized from 3,3'-dihydroxy-4,4'-diamino-biphenyl (HAB) and 2,2'-bis-(3,4-dicarboxyphenyl) Hexafluoropropane Dianhydride (6FDA). Polyimide films were converted to polybenzoxazole (PBO) by heating at 450 °C for 30 minutes. Under these conditions, nearly 100% conversion to PBO occurred. Chemical resistance of the PBO was investigated by exposing film samples to $50\%_{wt}$ ethanol/water at 120 °C and 3 atm for two weeks. Light gas permeability and selectivity was measured before and after exposure, and FTIR-ATR was used to probe changes to the chemical structure. Samples of the polyimide precursor exposed in this way were severely degraded, but fully converted PBO samples were little affected. Permeation tests involving water and ethanol vapors were performed at temperatures ranging from 110-190 °C and at pressures up to 6.4 atm. At 110 °C, water vapor permeability exceeded 3000 Barrer, and water/ethanol selectivity exceeded 60. Water vapor permeability was found to decrease with increasing temperature, indicating that water sorption was affected more strongly by the changing temperature than was diffusivity. On the other hand, ethanol permeability changed very little with temperature, indicating that the changes to sorption and diffusivity were more balanced for ethanol. For elevated pressures, mild plasticization effects were apparent, causing permeability to increase and selectivity to decrease with pressure.

(784c) Synthesis and Gas Transport of Thermally Rearranged Polyimides for Applications in Gas Separations

Smith, Z. P., The University of Texas at Austin
Sanders, D. F., The University of Texas at Austin
Guo, R., University of Notre Dame
McGrath, J. E., Virginia Tech
Paul, D. R., The University of Texas at Austin
Freeman, B. D., The University of Texas at Austin
Originally presented on: 11/8/2013 9:20:00 - 9:45:00

Over the past several years, research on thermally rearranged (TR) polyimides has shown interesting separation properties for a number of gas separations such as CO_2/CH_4 and hydrogen separations. TR polymers are traditionally synthesized via a solid state reaction of polyimides that contain reactive groups *ortho*-functional to the polyimide diamine. The precursor polyimides can be synthesized via a number of pathways, including chemical and thermal imidization. Furthermore, polybenzoxazole structures similar to those formed from polyimides can be synthesized by polyamides. The scope of this work is address the differences in gas transport properties for H_2, N_2, O_2, CH_4, and CO_2 as they relate to polybenzoxazole backbone structure, *ortho*-position reactive group, and synthesis of polybenzoxazoles from either polyimides or from polyamides.

Three *ortho*-hydroxy polyimides were synthesized via thermal imidization. The diamines for synthesizing these polyimides were 2,2-bis (3-amino-4-hydroxyphenyl)-hexafluoropropane (APAF), 3,3'-dihydroxy-4,4'-diamino-biphenyl (HAB), and the dianhydrides were 2,2'-bis-(3,4-dicarboxyphenyl) hexafluoropropane dianhydride (6FDA), and 4,4'-oxydiphthalic anhydride (ODPA). From these monomers, HAB-6FDA, APAF-6FDA, and APAF-ODPA were synthesized and converted to TR polymers by heating at 450°C for 30min. Conversion under these conditions resulted in nearly complete thermal rearrangement as determined by sample mass loss. The permeation properties of the precursor polyimides and TR polymers were determined for H_2, N_2, O_2, CH_4, and CO_2. Out of all of the samples considered, the APAF-6FDA polyimide and the APAF-6FDA TR polymer had the most promising transport properties out of all of the polyimides and all of the TR polymers, respectively. These samples had permeabilities near or above the 2008 Robeson upper bound for CO_2/CH_4 and H_2/CH_4separation.

A critical analysis on the effect of transport properties as they relate to the *ortho*-position leaving group was evaluated for HAB-6FDA. An *ortho*-hydroxy HAB-6FDA sample was modified via an esterification reaction post-imidization. This esterification reaction added different reactive functional groups to the *ortho*-position of the polyimide. In addition to the *ortho*-hydroxy group, an acetate, propanoate, and pivalate group were added to the polyimide backbone structure. These samples were thermally rearranged to different extents of conversion, and gas permeation was tracked as these samples were converted. Samples with bulkier leaving groups showed more dramatic improvements to gas permeability than *hydroxyl*-functional polyimides.

Finally, a polyamide sample of APAF-6FDA was prepared for gas transport properties. This sample can be converted into a polybenzoxazole via a condensation reaction at 350°C for 60min. The APAF-6FDA polyamide had consistently lower permeability than the APAF-6FDA polyimide. Furthermore, the polybenzoxazole synthesized from the polyamide had consistently lower permeability than the polybenzoxazole synthesized from the polyimide. It is believed that these differences in transport properties relate to the non-equilibrium nature of polybenzoxazoles prepared from different synthetic pathways.

(784d) Composite Hollow Fiber Membranes for Post Combustion CO2 Capture

Rownaghi, A. A., Georgia Institute of Technology
Koros, W. J., Georgia Institute of Technology
Bhandari, D.,
Originally presented on: 11/8/2013 9:45:00 - 10:10:00

Composite Hollow Fiber Membranes for Post Combustion CO_2 Capture

Ali A. Rownaghi[1], William J. Koros[1], Dhaval Bhandari[2]

[1]School of Chemical and Biomolecular Engineering, 311 Ferst Dr. NW, Atlanta, GA 30332

[2]GE Global Research Center, 1 Research Cir, Niskayuna, NY 12309

Fabrication of composite hollow fiber membranes with a thin polymeric permselective layer coated on a highly porous substrate for gas separation applications is attractive to provide both high flux and selectivity. The selective layer and the highly porous substrate described in this work are made from a combination of organic/inorganic materials. High porosity polyamide-imide (PAI) hollow fiber substrates were fabricated and characterized by scanning electron microscopy. The porous substrates were then coated with poly[bis-(2-(2-methoxy) ethoxy)ethoxyphosphazene] (MEEP) to create a permselective skin layer. The fabricated membrane performance was verified by permeation tests with He, N_2 and CO_2 at 35 °C and 30 psi. The membranes were found to have CO_2/N_2 selectivity up to 37. The increase in CO_2/N_2 selectivity for some membranes was accompanied by an increase in CO_2 permeance.

Acknowledgment

The material described in the abstract is based upon work supported by the Department of Energy National energy Technology Laboratory (DOE-NETL) under award number DE-FE0007514.

Disclaimer

(784e) Exploring Light Gas Transport in Carbon Molecular Sieve Membranes Using Diffusion NMR

Mueller, R., University of Florida
Kanungo, R., University of Florida
Kiyono-Shimobe, M., Georgia Institute of Technology
Koros, W. J., Georgia Institute of Technology
Vasenkov, S., University of Florida
Originally presented on: 11/8/2013 10:10:00 - 10:35:00

Carbon molecular sieve (CMS) membranes, microporous membranes fabricated from pyrolysis of polymeric precursors, are promising materials for energy-efficient separations of light gases. The understanding and prediction of transport properties of CMS membranes from only macroscopic observations is a very challenging task due to the complex amorphous pore structure of CMS membranes. Here we report results of an application of a pulsed field gradient (PFG) NMR method for microscopic diffusion studies in CMS membranes. In our multinuclear PFG NMR studies we combine the advantages of using large magnetic field gradients and high magnetic field. This enables obtaining reliable diffusion data even under the conditions of small sorbate diffusivities and short NMR relaxation times. As a result, PFG NMR measurements could be performed for diffusion inside CMS membranes on microscopic length scales of displacements under well-defined, uniform sorption loading and temperature conditions. The results of these measurements will be presented and compared with the previously reported macroscopic diffusion data.

(784f) ZIF/Polyimide Asymmetric Mixed-Matrix Hollow Fiber Membranes With Significantly Enhanced Hydrocarbon Separation Performance

Zhang, C., Georgia Institute of Technology
Koros, W. J., Georgia Institute of Technology
Originally presented on: 11/8/2013 10:35:00 - 11:00:00

ZIF/Polyimide Asymmetric Mixed-Matrix Hollow Fiber Membranes with Significantly Enhanced Hydrocarbon Separation Performance

Chen Zhang, William J. Koros

School of Chemical and Biomolecular Engineering, Georgia Institute of Technology

Asymmetric hollow fibers are the most industrially practical geometry for gas separation membranes in terms of productivity and membrane packing efficiency. However, studies on polymer-based mixed-matrix membranes in the past two decades have been mostly limited to the geometry of films with only a few extended to the geometry of asymmetric hollow fibers due to many technical challenges. For those few examples[1-3] in which selectivity enhancements were observed in mixed-matrix hollow fiber membranes, defect-sealing post-treatments (e.g. PDMS coating) were required in most cases to realize the enhanced selectivity.

This study reports successful formation of high performance asymmetric mixed-matrix hollow fiber membranes using a zeolitic imidazolate framework (ZIF) and a high T_g (glass transition temperature) polyimide, in which a significant propylene/propane selectivity enhancement was realized without any defect-sealing post-treatments. Spinning dope compositions and spinning parameters were optimized to form mixed-matrix hollow fiber membranes with desirable properties using the dry-jet/wet-quench technique. Propylene/propane gas mixture permeation showed that adding ZIF molecular sieve particles (~17 wt%) in the hollow fiber sheath layer significantly enhanced propylene/propane selectivity over pure polyimide dense film membrane by over 70% while maintaining attractive propylene permeance. The intrinsic compatibility of hydrophobic ZIF particles with the polyimide matrix as well as the optimized spinning dope compositions were keys for success of this study.

To our best knowledge, this is the first study in the open literature that reports successful spinning of asymmetric mixed-matrix hollow fiber membranes showing remarkably enhanced gas separation selectivity without any defect-sealing post-treatments. The current study, therefore, has significant impacts on the development of polymer-based gas separation membranes. An application of ZIF/polyimide mixed-matrix hollow fiber membrane described in this study is to pre-purify the feed stream of C_3 splitters in olefin plants to reduce the energy cost for producing polymer-grade propylene.

References:

(1) Husain, S.; Koros, W. J.: Mixed matrix hollow fiber membranes made with modified HSSZ-13 zeolite in polyetherimide polymer matrix for gas separation. Journal of Membrane Science 2007, 288, 195-207.

(2) Dai, Y.; Johnson, J. R.; Karvan, O.; Sholl, D. S.; Koros, W. J.: Ultem®/ZIF-8 mixed matrix hollow fiber membranes for CO2/N2 separations. J. Membr. Sci. 2012, 401Â¨C402, 76-82.

(3) Ekiner, O. M.; Kulkarni, S. S.: Process for Making Hollow Fiber Mixed Matrix Membranes. US Patent 6663805, 2003.

(814a) Polyimide Membranes With High Fractional Free Volume for Gas Separations

Wiegand, J.,
Guo, R., University of Notre Dame
Originally presented on: 11/8/2013 12:30:00 - 12:55:00

Polyimide Membranes with High Fractional Free volume for Gas Separations

Jennifer Wiegand (jwiegan2@nd.edu), Ruilan Guo (rguo@nd.edu)

Department of Chemical and Biomolecular Engineering, University of Notre Dame

Notre Dame, IN 46556-5637

As being realized, the biggest challenge for these polymeric gas separation membranes is the trade-off relationship between gas permeability and selectivity defining the so-called "upper bound". Over the years, substantial research efforts have been directed to expanding the polymeric membrane performance envelope via designing novel polymer materials. Particularly, it has been demonstrated that polymer membranes featuring high fractional free volume, such as thermally rearranged (TR) polymers in our previous studies[1-4] and polymers of intrinsic microporosity (PIMs)[5-7], provide unprecedented combination of high gas permeability and high selectivity. Here we report a new series of high performance polyimide membranes with high fractional free volume as well as tunable microcavities that have great potential for fast and selective transport of small molecules. These novel polyimides were designed and derived via incorporating a unique set of triptycene components in their backbone structures, which are bulky, three-dimensional, shape persistent units and possess rich structural hierarchy and versatile chemistry possibilities allowing unique chain arrangement and organization mechanism to induce hierarchically controlled architectures in the resulting membranes. Molecular design, synthesis and characterization of these new polyimides will be presented with a focus on how to incorporate and tune the architecture of the microcavities via introduction of side bulky groups and/or adjusting the interlinking geometry. Preliminary pure gas permeation data (with H_2, N_2, O_2, CO_2, and CH_4) will be presented for these new polyimide membranes. The correlation between microscopic structures (primary backbone structure, chain packing efficiency, fractional free volume, etc.) of the triptycene-containing polyimides with macroscopic membrane gas transport properties will be discussed.

References:

1. R. Guo, D. F. Sanders, Z. P. Smith, B. D. Freeman, D. R. Paul and J. E. McGrath, *Journal of Materials Chemistry A*, 2013, **1**, 6063-6072.

2. R. Guo, D. F. Sanders, Z. P. Smith, B. D. Freeman, D. R. Paul and J. E. McGrath, *Journal of Materials Chemistry A*, 2013, **1**, 262-272.

3. D. F. Sanders, Z. P. Smith, C. P. Ribeiro, R. Guo, J. E. McGrath, D. R. Paul and B. D. Freeman, *J Membrane Sci*, 2012, **409-410**, 232-241.

4. Z. P. Smith, D. F. Sanders, C. P. Ribeiro, R. Guo, B. D. Freeman, D. R. Paul, J. E. McGrath and S. Swinnea, *J Membrane Sci*, 2012, **415-416**, 558-567.

5. P. M. Budd, E. S. Elabas, B. S. Ghanem, S. Makhseed, N. B. McKeown, K. J. Msayib, C. E. Tattershall and D. Wang, *Advanced Materials*, 2004, **16**, 456-459.

6. P. M. Budd, B. S. Ghanem, S. Makhseed, N. B. McKeown, K. J. Msayib and C. E. Tattershall, *Chem Commun*, 2004, 230-231.

7. P. M. Budd, K. J. Msayib, C. E. Tattershall, B. S. Ghanem, K. J. Reynolds, N. B. McKeown and D. Fritsch, *Journal of Membrane Science*, 2005, **251**, 263-269.

(814b) Membrane Gas Separation Processes: Improved Performances Through Unsteady Cyclic Operation

Wang, L., Laboratoire Reaction et Genie des Procedes LRGP- CNRS, Universite de Lorraine
Corriou, J. P., Laboratoire Reaction et Genie des Procedes LRGP- CNRS, Universite de Lorraine
Castel, C., Laboratoire Reaction et Genie des Procedes LRGP- CNRS, Universite de Lorraine
Belaissaoui, B., Laboratoire Reaction et Genie des Procedes LRGP- CNRS, Universite de Lorraine
Favre, E., Laboratoire Reaction et Genie des Procedes LRGP- CNRS, Universite de Lorraine
Originally presented on: 11/8/2013 12:55:00 - 13:20:00

Membrane processes based on dense polymeric materials offer attractive performances for gas separation applications at industrial scale such as air separation, hydrogen purification, natural gas treatment or volatile organic compounds recovery [1].

Steady state operation is systematically performed for industrial installations and the separation performances are primarily governed by the membrane selectivity (a), which corresponds to the ratio of the compounds permeability through the polymer. In several cases however, the separation specifications cannot be reached due to too low membrane selectivity. In that event, the unique separation performances which can be obtained in some cases when a transient regime is applied can be of interest, especially when the differences between the diffusion coefficients of the permeating species largely exceed the differences in permeability. Paul explored this issue in a pioneering study, based on a cyclic process with synchronous valve operation [2]. We recently extended the analysis thanks to numerical simulation methods and optimization techniques which offer the possibility to explore asynchronous operation modes [3]. The concept proposed by Paul shows however a major limitation. Because the upstream composition has to remain unchanged, an infinitely low stage cut has to be applied (i.e. a very low fraction of the feed permeates through the membrane); as a consequence, the recovery ratio of the module is infinitely small.

In this study, a different approach, based on unsteady cyclic operation over long time scale (i.e. much larger than the time-lag of the system), a completely unexplored research area, will be detailed and described. The evolution of the upstream and downstream compartments pressure of a membrane module under cyclic unsteady operation have been first simulated through a tailor made computer program. The predictions obtained by simulation have been experimentally validated on a dedicated installation equipped with a hollow fiber gas separation module for air separation application (PPO fibers, Piccolo module from Parker). In a second step, the set of operating conditions which maximize the separation performances of the unsteady process have been identified through optimization. It will be shown for the first time that unsteady operation offers the opportunity to improve separation selectivity and overall recovery compared to steady operation [4].

The results of this exploratory study suggest to more systematically investigate the possibilities and limitations of gas separation modules under transient operation. A series of potential target applications (gas pairs) of industrial interest will be detailed. The possibilities of improvement of the concept and future challenges will finally be addressed.

References

[1] Baker, R, *Membrane Technology and Applications*, 2nd ed., John Wiley & Sons, New York, 2004

[2] Paul, D.R. "Membrane separation of gases using steady cyclic operation." *Ind. Eng. Chem. Process Des. Dev*, 1971: 375-379.

[3] Corriou, J.P., C. Fonteix, et E. Favre. «Optimization of a pulsed operation of gas separation by membrane.» *AIChE Journal* 54, n° 5 (2008): 1224-1234.

[4] Favre, E., Castel, C., Corriou, J.P., Wang, L. " Procede de separation membranaire en regime discontinu" French patent # FR 11 60587.

(814c) Multiwalled Carbon Nanotube Mixed-Matrix Membranes Containing Amines for Acid Gas Removal From Natural Gas
Ansaloni, L., Alma Mater Studiorum - Universite di Bologna
Zhao, Y., The Ohio State University
Ramasubramanian, K., The Ohio State University
Giacinti Baschetti, M., Alma Mater Studiorum - Universite di Bologna
Ho, W. S. W., The Ohio State University
Originally presented on: 11/8/2013 13:20:00 - 13:45:00

Multiwalled Carbon Nanotube Mixed-Matrix Membranes Containing Amines for Acid Gas Removal from Natural Gas

Luca Ansaloni[1], Yanan Zhao[2], Kartik Ramasubramanian[2], Marco Giacinti Baschetti[1], W.S. Winston Ho[2,3]

[1]Dipartimento di Ingegneria Civile, Chimica, Ambientale e dei Materiali, Alma Mater Studiorum - UniversitÃ di Bologna, Bologna, Italy

[2]William G. Lowrie Department of Chemical and Biomolecular Engineering, The Ohio State University, Columbus, Ohio 43210, United States

[3]Department of Materials Science and Engineering, The Ohio State University, Columbus, Ohio 43210, United States

Nowadays, natural gas is an important source for energy production.Â Depending on the original source, raw natural gas contains various impurities in composition, which must be removed before delivery to the pipelines. Â Removal of acid gases, namely, CO_2 and H_2S, from natural gas is therefore an important industrial gas separation process.

For CO_2 removal from natural gas, polymeric membranes compete against conventional separation processes, such as amine absorption, due to superior advantages of lower costs, lower energy consumption, and lower waste production.[1]Â Several attempts have been made to develop CO_2/CH_4 separation membranes that match the pipeline specifications. Â Glassy polymers, in particular polyimides, have been reported to show the best performance for this specific separation.[2,3] Â The higher condensability and smaller kinetic size make CO_2 molecules more soluble and diffuse faster in the membrane matrix than methane, resulting in good selectivity and permeability. Recently, polyimide-based polymer with intrinsic microporosity (PIM)[4] and cyclodextrine grafted co-polyimides[5] demonstrated very high CO_2 permeability at high pressure (10 atm) but low CO_2/CH_4 selectivity, less than 25. In fact, at high pressures, these membranes suffer considerably from the plasticization problem caused by high concentration of the acid gas, resulting in decreasing membrane selectivity significantly.Â Crosslinking can help to minimize the CO_2-induced swelling; however, an increase in separation factor of a solution-diffusion polymeric membrane always accompanies a reduction in the penetrant flow due to the well-known trade-off relationship.[6]

Facilitated transport membranes hold great potential for natural gas CO_2 removal. This type of membrane is able to achieve high CO_2 permeability accompanied with high selectivities vs. CH_4, H_2, or N_2 due to the specific, prompt reaction between amine carriers and CO_2 molecules.[7,8] This work exploited the superlative mechanical property of multi-walled carbon nanotubes (MWNTs) to reinforce the mechanical strength of amine facilitated transport membranes, from which we successfully synthesized nanostructured polymer-MWNT membranes by incorporating MWNTs (about 18 nm diameter by 180 nm) into the crosslinked polyvinylalcohol-polysiloxane network containing amine carriers.

The synthesized membranes inherited the facilitated transport mechanism and thus achieved high CO_2 permeability and selectivity vs. CH_4 at high pressures of at least 15 atm and high temperatures of more than 100Â°C. At 15 atm and 107Â°C using a feed gas of 50% CH_4 and 50% CO_2, which simulated the natural gas composition, the membrane showed a high CO_2/CH_4 selectivity of about 100 accompanied with a CO_2 permeability of about 350 Barrers. This lower permeability was due to the higher degree of carrier saturation at the higher CO_2 concentration. It should be noted that this permeability is still higher than that of commercial membrane, e.g., cellulose acetate (about 5 Barrers).

In this work, we also synthesized acid-treated MWNTs and amino-functionalized MWNTs as advanced nanofillers for amine facilitated transport membranes. The acid-treated MWNTs were functionalized with hydrophilic groups including carboxylic acid and hydroxyl groups by strong acid oxidation. The amino-functionalized MWNTs were chemically modified with 3-aminopropyltriethoxysilane.Â The presence of these hydrophilic groups was confirmed by FTIR spectroscopy. The membranes incorporated with the modified MWNTs demonstrated significant improvement in CO_2/CH_4 separation performance, especially in term of separation factor. A CO_2/CH_4 selectivity of 180 along with CO_2 permeability of around 1000 Barrers have been obtained at 15 atm and 107Â°C for a feed gas composition of 59.8% CH_4, 20.2% H_2 and 20% CO_2. These improvements are attributed to the modification of MWNTs. Â With the hydrophilic surface groups, MWNTs were compatible with the network of crosslinked polyvinylalcohol and polyamines.Â Thereby, the MWNTs were able to insert between polymer chains and disrupted polymer chain packing more effectively, resulting in increased free volumes between polymer chains. Â In addition, the membrane performance at higher feed pressure up to 60 atm as well as the effects of membrane composition, temperature and membrane thickness, has been investigated.

Keywords: multiwalled carbon nanotube; mixed matrix membrane; facilitated transport; carbon dioxide separation; natural gas.

References:

1. Baker, R. W. & Lokhandwala, K. Natural Gas Processing with Membranes: An Overview. *Industrial & Engineering Chemistry Research* **47**, 2109?2121 (2008).

2. Wang, L. *et al.* Gas transport properties of 6FDA-TMPDA/MOCA copolyimides. *European Polymer Journal* **44**, 225?232 (2008).

3. Wind, J. D., Paul, D. R. & Koros, W. J. Natural gas permeation in polyimide membranes. *Journal of Membrane Science* **228**, 227?236 (2004).

4. Ghanem, B. S., McKeown, N. B., Budd, P. M., Selbie, J. D. & Fritsch, D. High-Performance Membranes from Polyimides with Intrinsic Microporosity. *Advanced Materials* **20**, 2766?2771 (2008).

5. Askari, M., Xiao, Y., Li, P. & Chung, T.-S. Natural gas purification and olefin/paraffin separation using cross-linkable 6FDA-Durene/DABA co-polyimides grafted with α, β, and γ-cyclodextrin. *Journal of Membrane Science* **390-391**, 141?151 (2012).

6. Robeson, L. M. The upper bound revisited. *Journal of Membrane Science* **320**, 390?400 (2008).

7. Bai, H. & Ho, W. S. W. Carbon dioxide-selective membranes for high-pressure synthesis gas purification. *Industrial & Engineering Chemistry Research* **50**, 12152?12161 (2011).

8. Xing, R. & Ho, W. S. W. Crosslinked polyvinylalcohol?polysiloxane/fumed silica mixed matrix membranes containing amines for CO2/H2 separation. *Journal of Membrane Science* **367**, 91?102 (2011).

(814d) CO2 Removal From High Pressure Natural Gas With Hybrid Fixed-Site-Carrier Membranes: Membrane Material Development

He, X., Norwegian University of Science and Technology
Kim, T. J., SINTEF
Uddin, M. W.,
Hagg, M. B., NTNU

Originally presented on: 11/8/2013 13:45:00 - 14:10:00

CO_2 removal from natural gas (NG) is mandatory to meet the natural gas network grid specifications as CO_2 reduces the heating value of natural gas, is corrosive, and easily forms hydrates that can clog the equipment or damage the pumps. Processing of natural gas presents a large industrial gas separation market owing to the increased worldwide natural gas consumption. Traditional chemical (amine) absorption is well known and still considered as the state-of-the-art technology. However, membrane systems show a great potential for natural gas sweetening, even though it has only 5 % of the market today. Recently, strong interest has been focused on the development of different materials, especially for the composite membranes with a very thin selective layer, to overcome the challenges of the commercial membranes in natural gas sweetening. The novel fixed-site-carrier (FSC) membranes showed a good performance for CO_2 / CH_4 separation in our previous work [1-3]. However, high pressure still challenging the membrane operation due to the membrane plasticization and compaction caused by the strong adsorption of CO_2 and heavy hydrocarbon (BTEX) in the polymer matrix. The feasible solutions are the crosslinking of the membrane materials and development of pressurization-resistant membranes. Thus, the carbon nanotubes (CNTs) reinforced polyvinyl amine (PVAm) / polyvinyl alcohol (PVA) blend FSC membranes were prepared using an optimized membrane condition in current work.

The casting solution was prepared by adding a certain amount of CNTs into the PVAm (MW 340K) /PVA (72K) aqueous solution, and completely mixed using an ultrasonic mixer. The casting solution was evenly casted on the top of the polysulfone (PSf) supports following the procedure reported by Kim et al. [4]. The membrane structure and morphology was characterized by scanning electron microscopy (SEM), and a typical selective layer thickness was found to be ~1 μm. The membrane separation performance was characterized by gas permeation testing at 30 Â°C with a 10 % CO_2 - 90% CH_4 and a controlled feed flow rate. A lab-scale plate-and-frame module (membrane area 19.6 cm^2) and a small pilot-scale module involves 3 sheets (each sheet is 110 cm^2) were tested in high pressure permeation rig.

The permeation testing results indicated that the membrane preparation parameters such as heat treatment temperature, duration time and the pH of casting solution will significantly influence the membrane performance, and an optimal preparation condition is found to be: 95 °C - 0.75h ? pH10. Moreover, the operating parameters such as feed pressure, feed CO_2 concentration, and feed flow rate were also found to influence the membrane separation performance, which should be optimized in a specific application. The experimental results indicated **that CO_2 permeance decreases with the increase of feed pressure, which may due to the less contribution from facilitated transport mechanism (carrier saturation and low water vapor content at high pressure) and membrane compaction. It was also found** that both CO_2 permeance and CO_2/CH_4 selectivity are lower when the feed gas contains a high concentration of CO_2 50 % compared to a low concentration CO_2 at the same feed pressure. In addition, a high feed flow rate presents a much high separation performance due to a better flow pattern, but a relatively lower stage-cut. Therefore, a pilot-scale membrane module with larger membrane area should be tested in future work to document the membrane performance at a relatively high stage-cut which is usually involved in an industrial process.

The dependence of the membrane separation performance on the membrane area was tested with the small pilot-scale module, and the results indicated that both CO_2 permeance and CO_2/CH_4 selectivity decrease with the increase of membrane area, large membrane area gives a high CH_4 purity in the retentate but high CH_4 losses as well. Thus, a two-stage or multi-stage membrane system may be needed to accomplish a specific separation requirement (i.e., high CH_4 purity and low CH_4 losses). The developed FSC membrane showed a nice potential application for CO_2 removal from high pressure natural gas, but the process configuration and operating condition should be well designed and controlled in a real process. The current work can be well used to guide the process simulation to evaluate the process feasibility of CO_2 removal from high pressure natural gas process.

Acknowledgements

Â The authors acknowledge the NaGaMa project (partners: Norwegian Research Council, Statoil and Petrobras) for the funding of this work. The high pressure pilot-scale module designed by PHILOS is also acknowledged. The authors also thank SHOWA DENKO K. K. company in Japan to provide the carbon nanotubes for this work.

References

[1] Deng L, Kim T-J, Sandru M, Hagg M-B. PVA/PVAm Blend FSC Membrane for Natural Gas Sweetening. Proceedings of the 1st Annual Gas Processing Symposium. Doha, Qatar; p. 247-55.

[2] Washim Uddin M, Hagg M-B. Natural gas sweetening?the effect on CO2?CH4 separation after exposing a facilitated transport membrane to hydrogen sulfide and higher hydrocarbons. J Membr Sci. 2012;423?424(0):143-9.

[3] He X, Hagg MB. Hybrid Fixed-site-carrier Membranes for CO_2/CH_4 Separation. Euromembrane 2012. London, UK.

Â [4] Kim T-J, VrÃ¥lstad H, Sandru M, Hagg M-B. Separation performance of PVAm composite membrane for CO2 capture at various pH levels. J Membr Sci. 2013;428(0):218-24.

Â ADDIN EN.REFLIST

(814e) Performance and Stability of RTIL-Membranes for Dehumidification of Methane

Sullivan-Gonzalez, F.,
Amos, R., University of Mississippi
Bae, S. K.,
Scovazzo, P., University of Mississippi
Originally presented on: 11/8/2013 14:10:00 - 14:35:00

The biogas production from organic waste—animal, sewage, landfill material - could be a valuable renewable fuel. Before this biogas—predominately bio-methane—can be converted to fuel, it must be dehumidified. Typically membrane-based gas dehumidification uses hydrophilic polymer membrane materials, such as polydimethylsiloxane (PDMS) and cellulose acetate (CA). The main problems with polymer dehumidification membranes are methane loss and the susceptibility of polymer materials to plasticization by H_2O. Room temperature ionic liquid membranes (RTIL-membranes), according to recent literature, are not plasticized by water and have large water/CH_4 selectivities. In our paper, we will build on the initial literature RTIL-dehumidification data by looking at a series of ammonium and imidazolium-based ionic liquids. One membrane based on the trifluoromethanesulfonate [TfO] anion had a water permeance of 2000 GPUs with a H_2O/CH_4 selectivity of 12 000. With such high water permeances, boundary layer and support structures might significantly contribute to the overall water transport resistance, we will, therefore, present our work to measure and minimize this effect in our data. Of greater interest to the membrane community, however, will be our finding that there is an impact on some of the RTIL-membranes from the water vapor in the feed. This impact appears to be an increase in membrane mechanical stability and H_2O/CH_4 selectivity with increasing feed gas relative humidity. In an attempt to explain these phenomena we will explore the impact of the Feed rH the on surface tension of the stabilized RTIL.

(814f) Polymerized Pyrrolidinium-Based Ionic Liquid Membranes for CO2 Capture

Wang, J.,
Baker, S. N.,
Baker, G. A., Oak Ridge National Laboratory
Originally presented on: 11/8/2013 14:35:00 - 15:00:00

Ionic liquids (ILs) are salts with melting points below 100°C. They have been utilized for CO_2 capture due to some of their unique properties like high viscosity, low vapor pressure, and thermal and chemical stability. Utilizing IL membranes for CO_2 capture highlights the both properties of ILs and advantages of membrane technologies. Currently, supported IL membranes (SILMs) are playing a major role in the area of CO_2 capture. Such SLIMs have been fabricated with supported gelled and polymerized ILs. In SILMs, the use of porous supporting membrane materials, such as PVDF membranes or other organic frameworks, decreases the contact area of CO_2 and the effective IL components, which leads to lower capture efficiency. In this work, we are reporting IL membranes forming using only polymerized ILs without support materials. These membranes are then characterized and studied for their ability to capture CO_2 and to better understand the separation mechanism especially in terms of the cation and anion components.

AUTHOR INDEX

AUTHOR INDEX